CROOKED
MINDS

CROOKED MINDS

CREATING AN INNOVATIVE SOCIETY

KIRAN KARNIK

RUPA

Published by
Rupa Publications India Pvt. Ltd 2017
7/16, Ansari Road, Daryaganj
New Delhi 110002

Sales centres:
Allahabad Bengaluru Chennai
Hyderabad Jaipur Kathmandu
Kolkata Mumbai

Copyright © Kiran Karnik 2017

The views and opinions expressed in this book are the author's
own and the facts are as reported by him which have
been verified to the extent possible, and the publishers
are not in any way liable for the same.

ISBN: 978-81-291-4452-2

First impression 2017

10 9 8 7 6 5 4 3 2 1

The moral right of the author has been asserted.

Typeset by SÜRYA, New Delhi

CONTENTS

AUTHOR'S NOTE   vii

1. THE BLIND MEN AND THE ELEPHANT   1

2. ALL OF THE ABOVE...   25

3. IDEAS TO BENEFITS   54

4. WHERE DOES INDIA STAND?   76

5. CREATING INNOVATION HUBS   97

6. INDIA'S MOST INNOVATIVE CITIES   123

7. FOSTERING INNOVATION IN ORGANIZATIONS   161

8. A THOUSAND FLOWERS BLOOMING   184

AUTHOR'S NOTE

Buzzwords tend to have a short half-life; unlike most decaying radioactive elements, they reach half their active life in a fairly short period. Once-popular catchwords and phrases like 'core competence' are a rarity now. In this context, 'innovation'—a much-used word for many years—continues to retain its appeal. Beginning its journey mainly in the corporate world in developed countries, it travelled quickly to companies in developing countries, then to governments and also to the investment and grant-making community, and even to the world of art. Its longevity and appeal are probably indicative of it having real substance, rather than merely being the flavour of the month.

In the last few years, beginning at around the time when the Tata Nano caught the world's imagination, India is perceived as a global hotspot for innovation. Its offshoot, 'frugal engineering', has entered the vocabulary of many, particularly in the higher echelons of corporate management—not just in its literal application in the engineering industry, but equally in a metaphorical sense as a design or business concept. Stories about the products it engenders in India—from low-cost portable ECG and ultrasound machines to low-cost artificial limbs—have added to the country's reputation for innovation. Little wonder that global and domestic companies, in sectors from aerospace and automobiles to zoology and zinc, have set up centres for innovation and research and development (R&D) in India.

Apart from frugal engineering, and preceding it, is the idea of 'jugaad'. A major reason for India's standing as a centre of innovation is its jugaad culture—the innate capability to improvise, to find an out-of-the-box solution to almost any problem. Many regard this as innovation. However, most jugaad is hardly truly innovative; typically, it is just make-do or workaround solutions that are often not scalable. So, if jugaad is not it, what exactly is innovation?

At the other end of the spectrum, there are some who equate innovation with invention, with creating something new and therefore patentable. Yet, most would agree that a very different use of an already existing product is also an innovation, as is a new business model.

This book was born out of these thoughts, and my interest in trying to spot innovation in day-to-day life, rather than only in laboratories. Equally, in looking for innovation in art, films and literature; even in public policy. Spotting and highlighting such innovation is challenging, but a bigger driver has been in identifying the factors that make for a conducive environment or ecosystem for innovation. These include government policy, social norms, organizational structure, education, rewards and incentives, demographic or employee profile, etc.

The book covers the points mentioned above, beginning with why innovation is important, its definition and the systemic promoters (or obstacles) to innovation. It looks at where India stands, compared to other countries and the factors that account for its low rank. It also looks at some states and cities in India, and ranks the cities for innovativeness and future potential. Also discussed is what countries and organizations can do to encourage innovation.

The book picks up examples of innovation from a wide number of fields and from countries around the world. It is not a dissertation on innovation; accordingly, it has used newspapers, magazines and personal interaction as sources, rather than the many excellent academic studies and books on the topic. I have tried to write in a style that is direct, simple and—hopefully—engaging. The aim is to make the book accessible and interesting to a wide range of readers, including policymakers, corporate leaders, civil-society organizations, academic administrators and, indeed, anyone interested in innovation. I hope youngsters, particularly students, will not only enjoy it, but that it might help to trigger their latent innovativeness.

This book has seen the light of day only because of the help and support of a few people. Its birth owes much to Ritu Vajpeyi-Mohan, who first suggested, some years ago, that I write on innovation. She persisted with the idea through her own job transitions, and persuaded me that I should do this. She helped formulate the structure of the book and oversaw its progress (with great patience and occasional gentle prodding!) on behalf of Rupa Publications. The editing of the final manuscript was done, meticulously and with great thoroughness, by Amrita Mukerji. I am grateful to Ritu and Amrita for their assistance and dedicated effort.

I must acknowledge the ready help and input provided by many practitioners, quite a few of whom gladly spared time for discussions, besides providing specific information and inputs. Some of them have been mentioned in the book; others will remain anonymous.

Given the many other commitments that kept cropping up and my natural flair for procrastination, the book may have taken many, many months more, but for the constant pushing, encouragement and goading of my wife, Sunitee. She made sure that I was reminded of missed deadlines, and that I felt guilty enough to work harder!

Finally, by far the biggest contributor to the completion of the book was my daughter, Ketaki. She acted as my researcher, typist, proofreader, critic and more—all rolled into one. She was, simultaneously, my multi-tasking assistant and my supervisor. Despite her own commitments, she ungrudgingly spent long hours helping me, motivating me to get back to book-related tasks whenever I got diverted to other things. I doubt the book would ever have been completed without her help.

1

THE BLIND MEN AND THE ELEPHANT

What's innovation? Only something new: an invention? Or improvisation too? Why innovate and who does?

Jugaad and more examples—from outer space to down-to-earth.

Driverless cars, and those put together with bits and pieces of farm machinery; drones; self-healing aircraft; services provided profitably, but for which the user pays nothing; artworks that are based mainly on blank space; cities that reinvent themselves in new avatars; companies that reward failure: all examples of an exciting new world, based on innovations that have, and will, transform our lives. Innovation is now a key goal for organizations and governments, a sure recipe for growth, success and, often, survival. But what is innovation? A few examples may be the best way to start.

Some years ago, a farmer in rural India pondered over the problem of transportation. Like his fellow villagers, he often had to move farm produce and other goods to nearby villages, to the agricultural market, to towns, or between home and fields. Family and friends too needed to go from place to place.

Public transport was woefully inadequate and, in any case, did not reach his fields; nor could a car. A standard all-terrain vehicle would be far too expensive and would not survive the rough conditions for long. Faced with what seemed like an insurmountable problem with no solution, he came up with the idea of a 'home-made' contraption. With a thresher as its motive power, he added the necessary accessories and a trailer to carry people or goods, and created a new means of transportation.

Thus was born 'jugaad'—the vehicle; a word that has also come to mean improvisation. This has now become a generic tag for any such make-do, often out-of-the-box solution. India has attained much fame as a centre for such jugaad. Some look at improvisation as similar to innovation. But are these two—jugaad and improvisation—peas in a pod? A different product—again, a vehicle for transportation—provides an example of innovation. Google's autonomous or driverless car is based on dozens of innovations, beginning with its very concept—a car that needs no human guidance. To drive safely on busy roads, to obey traffic signals and rules, to stop where and when required—all these require a great number of new technologies, and also integration of existing technologies in new ways. The driverless car is now a proven concept. It will, of course, be some years before all the safety and regulatory issues are ironed out and it becomes just one more car on the road. Doubtless, its ultimate test will be on the chaotic roads of India, where the standard distance of vehicle separation is but a few inches, and stopping at a red light is not always the norm!

Airborne vehicles, which go from origin to a predefined destination, have been around for a long time (for example, missiles, beginning with the famous German V-2 rockets used in World War II and, in some sense, as far back as Tipu Sultan's rockets). In more recent times, unmanned aerial vehicles (UAVs) have been extensively used in conflict zones. Sometimes guided by pilots sitting (very much on earth) thousands of kilometres away, these drones are able to deliver bombs with pinpoint accuracy. The innovations required to do this are truly mind-boggling (even though they do not have to contend with rush-hour traffic on the Delhi–Gurgaon expressway as a driverless car may have to!). As interesting and innovative are miniature (almost insect-sized) drones used for surveillance.

From military uses, the concept of UAVs has spread to a growing spectrum of civilian uses. Drones for aerial photography are becoming common and they are now being experimented with for delivery of goods—from pizza to packages—by a number of companies. Unlike military UAVs, these necessarily require vertical take-off and landing capability (akin to a helicopter) and, therefore, innovations in a different area.

Sticking to transportation, there is another and more recent example from India.[1] Sajjad Ahmed, a school dropout from Bengaluru, has designed a non-polluting, solar-powered car. Named 'Surya Jyoti', it made a 3,000 km test run from Bengaluru to New Delhi. Put together in jugaad style, with parts from here and there, at a cost of Rs 1.2 lakh (less than US$ 2,000), it uses five solar panels to charge regular lead acid batteries. It can run for up to 40 km on full charge and at a top speed of 30 km per hour. The cost of running the car, according to the innovator, is Rs 1 per kilometre. The car

---

1. 'A Zero Pollution Solar Car: Possible Answer to India's Killer Air', Pallava Bagla, NDTV.com, 7 December 2015.

weighs about 450 kilogrammes and can ferry two passengers. Mr Ahmed is now travelling to Haridwar and then wants to head to Rameswaram to pay homage at the birthplace of former President Dr A.P.J. Abdul Kalam, his role model. That would be a 6,000 km trip. Such a car could be the answer to the dangerously high pollution levels in Indian cities. India's science minister, Dr Harsh Vardhan, took a ride in the car, and asked the Department of Science and Technology to support this effort so that the prototype can be made into a commercial success.

This solar car is probably somewhere between a simple improvisation and a real innovation, highlighting the difficulty in easy separation between the two.

Going back to aerial vehicles or aeroplanes, a recent breakthrough is undoubtedly truly innovative.[2] In work that the researchers themselves described as 'verging on science fiction', a team of British scientists has produced aircraft wings that can mend themselves after being damaged, indicating the tremendous potential of self-healing technology in the near future. The leader of the research team, Professor Duncan Wass, who has been working on the technology from 2012, said he expected self-healing products to reach consumers in the 'very near future'.

Professor Wass's team specializes in modifying carbon-fibre composite materials, the strong but lightweight substances used widely in the manufacture of commercial aircraft wings, sports racquets and high-performance bicycles. Their technology involves adding tiny hollow 'microspheres' to the carbon

2. What follows, on self-healing material, is substantially drawn from a piece by Chris Green in *The Independent*, London, 7 June 2015.

material—so small that collectively they look like a powder to the human eye—which break on impact, releasing a liquid healing agent. The agent seeps into the tiny cracks (which form in an aircraft's wings and fuselage) before coming into contact with a catalyst, triggering a rapid chemical reaction that causes it to harden.

Tests have established that the material is as strong after it has 'healed', raising the possibility of aircraft wings that can repair themselves literally on-the-fly if a bird strike takes place in mid-flight.

As in many other cases, this one too highlights the importance of drawing from several disciplines to innovatively solve a problem. In this instance, Professor Wass drew inspiration from biology. The human body, he noted, is 'not evolved to withstand any damage, but if we get damaged, we bleed, and it scabs and heals. We just put the same sort of function into a synthetic material'.

It seems clear from these and other examples that while some may easily be classified as either innovation or improvisation, others may lie in the grey area between the two. Unlike the clear distinction between discovery and invention, some may consider innovation and improvisation as similar, if not synonymous. While many differentiate between the two, and maintain that improvisation is not innovation, others argue that improvisation too is innovation. What, then, is innovation? The dictionary definitions of innovation, as spelt out in *The Oxford English Dictionary*[3], are:

- the action or process of innovating
- a new method, idea, product, etc.

---

3. *The Oxford English Dictionary*, Oxford University Press (first published 1 February 1884).

'Innovate' is defined as 'make changes in something established, especially by introducing new methods, ideas, or products'. Its origin is in the mid-sixteenth century, from the Latin *innovat*-'renewed, altered'; from the verb *innovare*, from in—'into' and *novare*—'make new' (from *novus*, 'new').

A more concise and practical definition—attributed to Schumpter—is: 'Innovation is the successful exploitation of a new idea.'

The dictionary definitions of improvisation, as spelt out in *The Oxford English Dictionary*[4], are:

- the action of improvising
- something that is improvised, in particular a piece of music, drama, etc. created spontaneously or without preparation

'Improvise' is defined as 'create and perform (music, drama, or verse) spontaneously or without preparation'. Apart from its use in the sphere of creative art, the dictionary also includes a broader definition, viz. 'produce or make (something) from whatever is available'.

Innovation might also be thought of as a means of getting more from less or fewer resources, for more people. Dr Mashelkar, former director-general of India's Council for Scientific and Industrial Research (CSIR), is amongst those who have popularized this concept of MLM, or more from less for more.

Innovation means different things to different people. It ranges from invention based on cutting-edge technological research to the makeshift jugaad vehicle; from new products

---

4. Ibid.

to improvisations in music. It has different facets, depending upon one's perspective, rather like the fable of the five blind men and the elephant, each of whom draws his own conclusion depending upon the part of the elephant he touches. Therefore, while one can get into an extensive academic debate on what constitutes innovation, it may be best to take a pragmatic view, as we do here, in which innovation is a broader concept and includes improvisation.

Innovation is certainly the new buzzword, the fashion of the day. It is a mantra repeated endlessly in companies, in academia, in books and even in the top echelons of governments. Boardrooms around the world echo to the ringing call to innovate, with questions as to what is being done to make the organization an innovator. But does innovation lead to concrete benefits? Why should individuals, organizations and governments innovate? After all, innovation is, by definition, something new, an idea that has not yet been tested. Is it not risky to abandon well-proven ways for a new, unknown and unproven approach? Business thrives on certainty, and innovation inevitably means a degree of uncertainty—moving off the familiar beaten path and on to a new, untraversed road.

These are indeed valid concerns. After all, much, if not all innovation is disruptive. On the other hand, that is in fact its strength—that it disrupts existing ways of doing things and can thereby give the innovator a special advantage over established players. The good thing about most innovations is that the consumer or user also benefits—through lower price, better quality, quicker delivery or an altogether new product or service (and sometimes through a combination of two or more of these factors). This makes the product or service more competitive. For incumbents in any field—the well-established major players—the threat from new and innovative companies is a constant worry, and one that is real and immediate today. If they are to continue thriving—or even to survive—their only response to such competitive threats can be to themselves innovate. It is also a means to expand their market base by reaching new customers. Little wonder, then, that companies look at innovation as the holy grail.

Companies that do not innovate, that do not seek new customers or markets, tend to fall by the wayside in due course. Researchers (Prof. Clayton Christensen, for example) have found that doing the right thing for too long can lead to failure (think of Kodak). As Christensen says[5]: 'Doing the right thing is the wrong thing.' Calling this the innovator's dilemma, he points out that this 'rears its head when…*disruptive technology* arises at the low end of the market.'

Companies now recognize the necessity to innovate in order to grow, to compete and, increasingly, even to survive. However, such an acknowledgement of the value and need for innovation is not limited to companies: countries, too, have realized the importance of innovation. Nations also are now getting on to the innovation bandwagon. A large number of countries now have ministries of innovation and active programmes to promote innovation. Countries—as also states and cities within them—seek to encourage innovation and compete to create the facilitative ecosystem that will attract innovators and spur innovation.

India formally declared 2010–20 as the decade of innovation and set up the National Innovation Council. Based

---

5. Clayton M. Christensen, *The Innovator's Dilemma*, Harper Business, 2000.

on its facilitation, many states in the country have created State Innovation Councils. Trade and industry bodies and academic institutions too have focused on innovation. Innovation funds have been started to provide early-stage capital to innovators, and there is a proposal to create Innovation Fellowships for students.

Many countries have taken initiatives to further innovation. One example is China. Faced with a slowing economy and an ageing population, China is counting on innovation to reinvigorate the nation. Prime Minister Li Keqiang, in a March 2015 report to the National People's Congress (Parliament), identified entrepreneurship and mass innovation as one of the two 'twin engines' (the other being increased supplies of public goods and services) that should drive the economy of the future.

Further, China's State Council, chaired by the prime minister, announced in 2015 that the country will pilot online crowd-funding, in addition to encouraging banks and financial institutions to provide loans to small businesses. The science and technology minister indicated that over a thousand investment agencies were channelling as much as US\$ 56.8 billion for promoting entrepreneurship. Start-up technology companies are being supported by 115 university science parks and over 1,600 technology business incubators, which are incubating over 80,000 enterprises.[6]

Another example is Norway, a developed and prosperous country. This nation presents an example of a country that plans to use innovation to secure its future. Thanks to its discovery of massive oil reserves in the 1960s, Norway is now amongst

---

6. These details were covered in a report in *The Hindu* (Delhi edition), 13 March 2015.

the richest countries in the world. Like its neighbours, it has a strong social security system, as also free education and high wages. Such a welfare state, though, requires robust ongoing revenues, especially in a country with an ageing population (thanks in part to an excellent and assured healthcare system) and therefore, high pension payouts.

Norway's ability to support the extensive welfare programmes is based on a strong economy, the underpinning of which is oil. Hydrocarbons account for over two-thirds of Norway's exports, a fact that highlights the country's dependency on oil. With its volatile price, revenues from oil are not easily predictable. Also, despite its large reserves, at some point the country may begin to run out of oil. What then? The government has prudently saved money, ensuring a large balance and also creating a sovereign wealth fund, which is well invested. Even so, its political and intellectual leaders are concerned and have been looking at alternatives.

One alternative, in place since 2004, is an organization called Innovation Norway.[7] It is a unique combination of trade council, business incubator, angel investor and promoter. This state-owned company is leading the effort to move the country towards a more diversified economy, one that will depend less on its oil and gas resources. Innovation Norway invests in various green technologies too, like wind energy. In 2013, various ministries funnelled half a billion dollars through Innovation Norway.

Here, then, is a country that sees its future being driven by innovation, and has created an organization dedicated to taking forward this vision.

7. Based on a report in *The Hindu*, 25 October 2014.

China and Norway are but two examples of what is increasingly a widespread trend. Innovation has become even more important with globalization, which has been spurred by the increasing ease of trade, facilitating the cross-border movement of goods and services. In parallel, the phenomenal progress—a revolution, one might say—in information and communication technology (ICT) has made possible the instantaneous flow of data, audio and video. Telephone conversations, television broadcasts and data of all kinds flow around the world with no obstacles. In this sphere, national boundaries have practically melted away and we live in a McLuhan-esque 'global village'.[8] In this seamless and borderless world, competitive advantage, whether for organizations or for countries, cannot be long sustained on the basis of wage or cost arbitrage. Ultimately, it must derive from innovations that provide greater value for money, more bang for the buck.

One example is India's IT industry. Its success and phenomenal growth can be traced to an innovative business model, based on 'outsourcing and offshoring'. In this, customers—particularly those in high-cost and talent-constrained countries like the US, the UK and Japan—outsource small or comprehensive work packages to IT companies in India. Such global supply chains, stretching across multiple countries, have long been common in manufacturing, but were rare for services. The ICT revolution provided the impetus for actually realizing the concept of 'remote delivery': services that could be rendered from a location far removed from the user or customer. This

---

8. Marshall McLuhan, *The Global Village*, Oxford University Press, 1968.

depends upon reliable and instantaneous transmission of data, at low cost, across continents. It was also facilitated by new technologies—both hardware and software—that made feasible such trans-shipping of work.

Paul Samuelson, Nobel laureate, is famously reported to have said, 'You cannot export a haircut.' One would have thought that this was equally true for secretarial services. Yet, today, it is not uncommon for executives to have their secretaries sitting thousands of kilometres away, instead of being a few feet down the corridor. After all, practically everything—except bringing you coffee!—that can be done by a secretary sitting in your office can be done by someone in a different continent. All it needs is an excellent communication network, good computer software and the usual office equipment. Such an innovation was put into practice by OfficeTiger, amongst others.

OfficeTiger was founded by Joseph 'Joe' Sigelman and Randolph 'Randy' Altschuler, both raised in New York City.[9] They first met when both were students at Princeton University and became good friends. Later, Joe and Randy went on to Harvard Business School together. In 1999, bitten by the entrepreneurial bug, both Joe and Randy resigned from their well-paying jobs as investment bankers, packed their suitcases and travelled—not to Silicon Valley, the Mecca for start-ups; not even to Bengaluru, but to Chennai. Joe's shocked mother reportedly asked him, 'Joe, you went to all these good schools and universities and got good pay. Now, you want to go to India and be unemployed?' At that time, except for General

---

9. Some details in the two paragraphs on OfficeTiger are from 'OfficeTiger Roams Towards an IPO', *Bloomberg*, 12 September 2005, and 'OfficeTiger: An Amazing Success Story', rediff.com, 14 September 2006.

Electric (GE), hardly any American company was doing any major outsourcing work from India; this was, therefore, a pioneering effort.

Indicative of the success of OfficeTiger's innovative effort is the fact that in April 2006, it was acquired by R.R. Donnelly & Sons for US$ 250 million. OfficeTiger then had 10,050 employees (with 4,000 in India), twenty-nine delivery centres and forty-two client sites across nine countries.

Innovation, as it is commonly understood, generally means new products or the leveraging of technology. Yet, a lot of it comes from a creative idea, a concept that can be transforming. Consider, as an example, Las Vegas. Founded in 1905, it was but a small town, serving as a resupply point for travellers going to California. Within a few years, water from wells was piped into the town, providing both a reliable source of fresh water and the means for additional growth. The availability of water in an otherwise arid area allowed Las Vegas to become a water stop, first for wagon trains and later for railroads, on the route between Los Angeles and destinations such as Albuquerque, New Mexico.

In 1950, Las Vegas had a population of 25,000; in 2016, its population is estimated at 625,000.[10] The transformation of the city is indicated by two other figures: US$ 9.6 billion, its gaming revenue in 2015; and 42 million, the number of visitors in the same year.[11] Its transition from a watering hole and wayside stop in the desert to an economic powerhouse is based on its conversion into a gaming centre, convention city and holiday destination. In recent years, it has also reinvented

10. Estimated by City of Las Vegas Redevelopment Agency.
11. Las Vegas Convention and Visitors Authority.

itself from an adult-oriented gambling centre to a family entertainment destination with many attractions for children as well. When in Las Vegas, one feels as if one is in a fairy tale or dream city. This is true, in a way; after all, it is a city that is built on a dream.

A similar example is Dubai, another story about how innovative ideas can metamorphose a city. Like Vegas, once a nondescript town in a desert, Dubai first became an oil boom town in an inhospitable environment: a place that no one visited unless they had work there. Dubai has now become a tourist hub, with more than 10 million visitors in 2012.[12] Its attractions include massive shopping facilities, with some of the largest malls in the world; a large indoor ski resort where one can ski while the outside temperature crosses 45^0C; an upcoming theme park by Universal Studios; and life-size replicas of the seven wonders of the world. It seeks to attract sports-tourists through F1 car racing, major golf and tennis tournaments and yachting. Its iconic buildings—like the Burj Khalifa (which gained the official title of 'Tallest Building in the World' at its opening on 4 January 2010)—and its artificial archipelago of small islands (4 km off the coast) are also big tourist draws. Dubai's airport is planned not only for traffic whose destination or origin is the city, but also as a major interchange and transit hub. In 2015, it had 78 million passengers going through, crowning Dubai as the world's busiest airport (surpassing Heathrow's 74.9 million passengers).[13] It expects 100 million passengers a year by 2020.

---

12. 'Dubai welcomed record-breaking 10 million tourists last year', *The National*, 7 March 2013.

13. 'Dubai International retains world's busiest airport title; handles 78 million passengers in 2015', *Gulf News*, 1 February 2016.

A large number of Indian travellers heading to Europe, the US or Africa use Dubai as the transit point. Dubai aims to serve as a global financial centre and become a knowledge hub (it has already attracted many foreign universities).

Dubai is now planning to build a temperature-controlled 'city', which will include the world's largest mall ('Mall of the World', covering 8 million square feet) and an indoor park, as well as hotels, health resorts and theatres.[14] The all-pedestrian complex will occupy a total area of 48 million square feet. The theme park will be under a glass dome that will be opened in winter. The promenades connecting the facilities, some of which are as long as 7 km, will be covered and air-conditioned during summer. Dubai hopes that this will make it an attractive tourist destination all around the year; it is expected that the mall will see 180 million visitors a year. Such creative and visionary thinking seeks to reduce Dubai's dependence on oil revenues, and it has already succeeded in repositioning itself.

Both Las Vegas and Dubai demonstrate the power of innovative thinking, of how vision, creativity and great execution can transform forlorn desert towns into prosperous economic centres.

Companies too have shown how innovation can turn around their fortunes. The example of Apple is well-known: a company that was on the verge of disaster in 1997, it is now amongst the most valuable companies in the world. On 20 August 2012, Apple became the most valuable public company in history, breaking the record for market capitalization that Microsoft had set in December 1999. Its market capitalization

---

14. Dubai Holdings media release, 5 July 2014.

reached a whopping US$ 624 billion.[15] In 2015, some analysts predicted that it will cross US$ 1 trillion in the next year or two.

Apple has a chequered history. Founded in 1976 by Steve Jobs, Steve Wozniack and Ronald Wayne (who sold his share soon after), its first product (Apple 1) was an assembled circuit board—sold without a keyboard, monitor or case—priced at US$ 666.66.[16] For the first five years, it is reported to have averaged a growth of 700 per cent. In December 1980, Apple went public at US$ 22 per share, instantly creating more millionaires (about three hundred) than any other company till then.[17]

After disputes within the company, Jobs left Apple in 1985. A decade later (in 1996), Apple acquired a company called NeXT and with it, reacquired Steve Jobs, who returned as an advisor. In 1997, after a three-year record-low stock price and crippling financial loss, the board removed the CEO, Gil Amelio, and appointed Steve Jobs as interim CEO. In 2001, Apple announced the iPod portable digital audio player. It was phenomenally successful, with 100 million sold in six years.[18] In 2003, the iTunes store was introduced, offering online music downloads for US$ 0.99 a song and integration with iPod. There were 5 billion downloads in the first five years.[19] These

---

15. 'Apple's market value exceeds Microsoft's during 1999 bubble', Reuters, 21 August 2012.

16. Computer History Museum.

17. 'From IBM to Apple: the good, the bad and the ugly of tech IPOs', The Telegraph, 18 May 2012.

18. 'An Apple milestone: 100 million iPods sold', NBC, 4 September 2007.

19. 'iTunes store: 5 billion songs; 50,000 movies per day', Fortune, 19 June 2008.

successes were followed by those of iPhone, Apple's AppStore, iPod Touch and iPad.

Doubtless, many factors contributed to its stunning turnaround and success, but certainly innovation was amongst the most important. Each of its products was an outstanding innovation in design, as much as in branding, positioning and overall marketing. It is interesting to note that none involved major technological breakthroughs, so the innovations were really in 'soft' areas. Apple's reputation—and its market valuation—is built as much on its present products and financial success, as it is on its acknowledged capabilities to innovate. An indicator of its reputation, built mainly on its innovative capability, is its brand value. In its annual report (2013), Interbrand—which compiles, each year, the Best Global Brands report—placed Apple as the first amongst the hundred most valuable brands, displacing the thirteen-year front runner Coca-Cola. The report estimated the value of the Apple brand at US$ 98.3 billion.[20]

The growing recognition of the importance of innovation is underscored by the fact that technology companies—known for innovation—occupied five of the top ten positions in the Interbrand report. Also in the top ten were two other companies—IBM and GE—widely acknowledged as innovative. Given this, it is not surprising that the poster child of innovation, Google, was a close number two, with a brand value of US$ 93 billion.

Concrete evidence of how innovation is rewarded comes from that ultimate arbiter—the market. Apple was mentioned earlier; another, and more recent example, is Twitter. It

---

20. Interbrand's 15th Annual Global Brands Report, 2014.

innovated the creation of a micro-blogging site, which allows posts of 140 characters. Twitter has grown rapidly; by the second quarter of 2015, it was averaging 316 million actual users a month.[21] It has become a phenomenon in itself, and many even credit it as being amongst the key mass mobilizers that drive revolutions (for example, the 'Arab Spring'). During 2015, its revenue was an impressive US$ 2.2 billion.[22] Its market capitalization, though lower than the peak of US$ 39.9 billion (December 2013[23]), was US$ 16.06 billion[24] at year-end 2015: really huge for such a young (and loss-making) company. What better proof can there be of the market rewarding innovation!

If more evidence is desired, it comes from examples even more recent than Twitter. The runaway success of Uber and Airbnb point to the extent of market recognition for innovation. Both companies basically provide technology platforms, based on the simple—though innovative—idea of monetizing services that tap into underutilized assets. Uber was valued at US$ 68 billion during its June 2016 round of fund raising. [25] Airbnb, a few months later (August 2016), was valued at US$ 30 billion.[26] Chapter 2 has more on both these companies.

---

21. Q2 Earnings Release from the Twitter website, 28 July 2015.

22. Statista: The Statistics Portal.

23. From YCharts, 21 August 2015.

24. Ibid.

25. 'Uber Raises $3.5 Billion From Saudi Fund', *The Wall Street Journal*, 1 June 2016.

26. 'Airbnb Files to Raise $850 Million at $30 Billion Valuation', *Bloomberg Technology*, 6 August 2016.

Though Apple, Twitter, Uber and Airbnb are all technology companies, innovation, as noted earlier, is not driven by technology alone; nor does it necessarily culminate in a product. A non-technological innovation that created a multi-generational impact comes from ancient China. Vikram Seth recounts this in the Foreword to his book, *The Poems 1981–1994*.[27] He tells us about the erstwhile concubine of Tai Zong, founder of the Tang dynasty in China: in the year AD 683, she installed herself on the throne (as Empress Wu) after the death of Tai Zong and his son. While her fifteen-year rule made her a controversial figure, one of her innovations was the inclusion of poetry as a compulsory subject in the imperial civil service examinations. This had a profound effect, and the resurgence of poetry in China at that time had much to do with this seemingly small change. Empress Wu may, therefore, deserve as much credit for Chinese poetry as the poets themselves!

In more recent times, a somewhat similar long-term impact resulted from an 'innovation' that was triggered by war. During World War II, the British felt a need to convey their perspective to Indians. This was probably the result of their concern about whether India's struggle for independence would result in Indians refusing to support the Allies in their war effort. Doubtless, this worry was exacerbated by Subhas Chandra Bose and his Indian National Army, which was willing to work with the Japanese to oust Britain from India. One of the ways Britain chose to counter anti-British sentiment was through propaganda films. To ensure that these would get maximum viewership, it was made mandatory for every cinema theatre

---

27. Vikram Seth, *The Poems 1981–1994*, Penguin Books Ltd, 1996.

in India to screen such a news-related documentary before the feature film, during every show. At the end of the war, and after India became independent, this compulsory screening of a short government-provided film was continued. Newsreels were produced by the government-run Films Division and screened at every show. Over time, these newsreels were supplemented with (non-news) documentary films—some produced by Films Division and others made by independent filmmakers. Given the law of compulsory screening, there was an assured and robust demand for such short films, and this resulted in a vibrant documentary production ecosystem, giving a huge fillip to documentary films and filmmakers. Thus, in much the same way that the diktat of a Chinese empress, over a millennium ago, gave rise to generations of poets, a British wartime law produced generations of documentary filmmakers in India.

Some of these examples do bring us back to the question, what is innovation? Can a new policy (as in the examples of Chinese poetry or documentary films in India) be termed an innovation? Does an innovation necessarily have to be something new? At least on the last question, a vast majority will probably answer in the affirmative. Yet, there are many cases where an existing product is put to a different use and most will acknowledge that it is truly innovative. An example is the story—doubtless hackneyed, but nevertheless worth recounting—from the American space programme. By all accounts, it has at least a kernel of truth, even if each raconteur embellishes it in his or her way (as is done here!).

In the early days of manned space flight, astronauts were required to note down various figures as part of scientific experiments or for monitoring their own temporary habitat in space. The then-standard fountain pen could not be used, as its ink would leak out in a low-pressure environment (as some from my generation experienced while in an aircraft, with the telltale embarrassing inkblot on their shirt pockets!). A ballpoint pen, though quite safe to carry on an aircraft, could not be used in the low-gravity environment of space, as its functioning is based on ink flow due to gravity. Therefore, NASA—in its usual meticulous style—put out a request for proposals and commissioned the development of a pen that could write in low-gravity situations. The result, a marvel of technology, is the 'space pen', also known as the zero gravity pen, marketed by Fisher Space Pens. It was originally used in Apollo 7 in 1968, and can be seen at the Smithsonian Air and Space Museum in Washington DC.

The Soviets, faced with the same problem in their manned space mission, came up with a different solution (as the story goes)—they used a pencil!

This is certainly a dramatic example of innovation. The real story is, unfortunately, less dramatic.[28] It is apparently true that NASA intended to develop a space pen, but when estimates of the development cost skyrocketed, the project was abandoned.

Initially, both NASA astronauts and their Soviet counterparts used pencils. However, it was realized that pencil tips which flaked or broke off would drift in the microgravity environment of space, posing potential harm to equipment or astronauts. Also, NASA wanted to avoid inflammable material, especially after the Apollo 1 fire.

---

28. Details of the space pen story are from 'Fact or Fiction?: NASA Spent Millions to Develop a Pen that Would Write in Space, whereas the Soviet Cosmonauts Used a Pencil', Ciara Curtin, *Scientific American*, December 2006.

In 1965, NASA ordered thirty-four mechanical pencils from a Houston company, and paid US$ 128.89 per pencil. When this became known, there was a public outcry and NASA began to look for a cheaper alternative.

In the same year (1965), Paul Fisher's company, the Fisher Pen Company, patented an 'antigravity' pen that could write in any orientation (even upside down) and at practically any temperature (from minus 50°F to 400°F)—or even underwater. The Fisher pen—unlike ballpoint pens—does not rely on gravity to get the ink flowing; instead, the cartridge is pressurized with nitrogen. This prevents air from mixing with 'ink'—which, too, is not standard ink—so it cannot evaporate or oxidize.

The myth that NASA spent millions on the development of the space pen is incorrect, and it was developed using private capital (Fisher reportedly invested US$ 1 million to develop it). After extensive tests, NASA decided to use the Fisher pens and ordered four hundred of them for the Apollo programme in 1968. A year later, the Soviet Union ordered a hundred pens and a thousand ink cartridges for their Soyuz missions. Interestingly, both paid only US$ 2.39 per pen (after a 40 per cent bulk purchase discount). Compare this with the reported price paid by NASA for the mechanical pencil (US$ 128.89) and one sees the tremendous potential of disruptive technological innovation.

Like the pencil, a similar simple solution to a pressing need came from India and again, from the space sector—though, in this case, the problem was on the ground. The year was 1975, and the Indo-US Satellite Instructional TV Experiment (SITE), after years of planning and preparation, was inaugurated in August. Hailed by renowned science fiction author Arthur C.

Clarke as 'the biggest communication experiment in history', SITE involved the use of a cutting-edge US/NASA satellite, ATS-F (one of a series of NASA's advanced technology satellites), to broadcast TV programmes to community sets in 2,400 remote villages in India. The television signals, beamed up to the satellite from ISRO facilities in Ahmedabad and Delhi, were received on the village TV sets through a dish antenna with a diameter of 3 metres. This 'direct broadcast', as it was then called, was thus a precursor to today's direct-to-home (DTH) TV services.

Keen to take the programmes to truly remote and deprived communities, ISRO installed many of the TV sets in villages without electricity. These receivers were powered by batteries, which were recharged periodically. The total daily broadcast was for only four hours, and the charging cycle, involving a trip to the village by ISRO engineers, was calculated on this basis. However, it was found that in a few villages at first, and in many later, the batteries were discharging much faster. This foxed the ISRO engineers, who had carefully and painstakingly worked out the charging cycle based on theoretical calculations as well as actual laboratory tests. A faster discharge was a blow to their knowledge and skills! Conversations with the villagers finally solved the mystery (and restored the engineers' self-confidence)—the TV set in each village without electricity was kept on for many hours after the broadcast had ended, so as to serve as a tube light! This was used for reading or for a gossip session after the end of the broadcast. The pencil in space was an existing product being put to its intended use (writing), but in a different context (in outer space); here, an existing product was being put to an unintended use (as a source of light) because of the context (no electricity in the village).

These stories help to illustrate two very different facets of innovation, practically at extreme and opposite ends of the innovation spectrum. The 'pencil solution' to the problem (though not fully factual) was simple, direct, quick and inexpensive. The Indian villagers too met their need through a solution that was quick and direct. Both were certainly innovative—and in retrospect, obvious. These are, in fact, characteristics shared by many innovations. At the other extreme are those that involve painstaking, time-consuming and expensive research and development (R&D). Thus, innovations span a very wide range on all dimensions: time, cost, complexity, effort and impact, amongst others. Also, as seen in the few examples outlined earlier, they include various categories: innovations in product, policy, process and business model. It can mean existing needs met by a new product, an existing product meeting a new need, an existing need being met by an existing product used in a new way, or a new product meeting a new need or want. Thus, as noted earlier, innovation—at least as we define it in this book—is not just one aspect or another of the above; rather it is, as further elaborated in the eponymously named next chapter, 'all of the above'.

The examples given here answer, at least in part and somewhat tangentially, the questions we began with: what is innovation, and why do we innovate. In the chapters that follow, some of the aspects mentioned here are covered in greater detail. Also discussed are the factors that drive and sustain innovation; how innovation can be promoted; where India stands compared with other countries; and the pecking order amongst different cities and states within India.

2

ALL OF THE ABOVE...

The many facets and areas of innovation: business models, products, processes; in art and literature; in government and the social sector. Disruptive ideas that create public gain or private profit. Crooked thinking and creative cheating.

In the late 1960s, washing machines were new to India. They were also very rare, given the cost and other constraints. Therefore, when a friend—newly minted from a top management school—was given the task of selling washing machines, he was hardly the subject of much envy. For many months, whenever we met, he would crib about his job and drown his grief in endless cups of coffee. After all, sales and marketing of consumer durables was fine, but to sell washing machines in India was impossible, he announced. How could one sell washing machines when Calcutta had no power, homes in Bombay had no space, Madras had no water and in Delhi everybody had household help? Obviously, the market for a product like this—which offered convenience and added status—would be mainly in the four big metros.

One fine day, our glum and depressed young man was all smiles and offered to pay for everyone's coffee. The reason?

He was way over his target for washing-machine sales in the quarter. What suddenly changed, we all wondered. He was equally mystified at the great turnaround, and even more so by the fact that most of the sales were not in the metros, but in Punjab. Now, who would want to buy washing machines in mainly rural Punjab, even if it was relatively prosperous? No one had an answer, but as our friend said, why look a gift horse in the mouth—sales were sales.

Yet, being a fired-up professional with classroom exhortations still ringing in his ears, our friend embarked on a market-research mission to profile the buyers and see if the lessons could be translated to increase sales in other parts of the country.

Our next coffee session was like the final chapter of an Agatha Christie mystery: the unexpected solution is spelt out and then seems so obvious that you wonder how you did not see it. In this case it was—as is well-known through endlessly repeated anecdotes—that the washing machines were being bought in Punjab, not to wash clothes, but as giant lassi-churners! So, when the typical gang of friends (or large family) turned up at the local dhaba or restaurant and ordered ten glasses (all super-size) of lassi, the chef did not have to painfully produce them one glass at a time in his 'mixie', but was able to turn out gallons of it in one go. Who thought of this first remains a mystery (notwithstanding Agatha Christie), though my friend would, of course, like to claim full credit. However, once the proof of concept was established, the idea spread like wildfire (and my friend shattered all his targets).

This is but one of the countless examples of innovating by using an existing product for a very different purpose or—as seen in the pencil-in-space case—using it in a context for which it had not been made. Such innovation demands creative thinking; thinking that is not constrained by the rigidities of the product-use framework. Most people are trapped in the routine acceptance of a product having been created for a defined use. It is only the rare innovative mind that looks at the inherent capabilities of the product and sees where else or how else it might be used. Sometimes, the innovation begins not with the product but with the problem (for example, how to take notes in a low-gravity environment). Typically, the problem too needs to be thought of at a level deeper than the superficial. Thus, it is not 'how to use a car on rough or non-existent rural roads', but 'how to transport people or goods in a rural area'. Therefore, while one innovative strand may be the design of a truly all-terrain vehicle (like the US Apollo Lunar Roving Vehicle or, more recently, China's moon rover, Yutu), another is the use of existing products to create the jugaad transporter. While the approach of using an existing product is quicker and cost-effective, large and long-term benefits generally come from new inventions or products.

This is most visible in the field of technology, where path-breaking new ideas have been translated into products that have not only changed our lifestyle, but have often led to sustained economic growth. The printing press, the steam engine, telephony, transistors, micro-electronics and computers are but a few examples of inventions that have revolutionized life and economic growth. Therefore, though many of the innovations discussed here are not in the category of 'inventions', it is necessary to emphasize the critical importance of such altogether new products or technologies. In fact, this is the reason why most lists of innovative companies or countries use 'number of patents' as a criterion. Many also use 'R&D

expenditure' as a surrogate or as a lead indicator for patents and inventions.

Invention, though, needs to be translated into a viable product for social or economic benefit. This requires an appropriate business model or—for innovations aimed at social but not economic gain—a process of production–distribution–delivery combined with marketing or demand creation. In many cases, it is this part of the value chain that requires as much innovation as the invention itself. Equally, innovations in design—especially for consumer durables—can be of great importance. Apple products best exemplify this: the company's USP for all of them is not new technology, but innovations in design (and, of course, clever marketing).

A classic example of an innovative business model comes from the Blockbuster–Netflix story.[1] The convenience of watching whichever movie one wanted to, at a time of one's choosing and in the comfort of one's home, was a reality made possible by Digital Video Discs (DVDs). Blockbuster built its business around this, with stores all over the US, and virtually dominated the movie rental market in the 1990s. Its model required it to hold a large inventory of movies, entailing substantial investment. Revenues were determined by the turnover of this inventory—DVDs sitting on the shelves were a drag, while DVDs in the customer's home meant income. To maximize its return, it needed the customer to return the DVD quickly, so that it could then be rented out to the next customer. To ensure this, it charged a substantial fine for any delay in returning the DVD on time. Given the human tendency

---

1. This example draws generously from Clayton Christensen (with James Allworth and Karen Dillon), *How Will You Measure Your Life?*, HarperCollins Publishers, 2012 (pp. 179–183).

to procrastinate, there were always delays and, as its fines increased, Blockbuster was soon making most of its profits from such fees for late return.

Netflix began, in the latter half of the 1990s, with a model of mailing DVDs to customers rather than requiring them to go to a store. It charged a monthly fee, like a membership, rather than a per-movie charge. So, when a customer was too busy to watch a movie s/he had ordered, the DVD sitting in the customer's house (rather than in Netflix's warehouse) was not causing a loss; in fact, it was profitable for Netflix, as there was no need to pay for the return postage or to send out the next batch of movies for which the customer had already paid (through the monthly fee).

This was a very different business model from that of Blockbuster, and one that required innovative thinking. Blockbuster was huge and dominant—billions in assets, stores across the country, thousands of employees and instant brand recognition. Even in 2002, it was confident enough to assert that 'Online rental services are serving a niche market'. A decade later, Netflix had an estimated 24 million customers and Blockbuster had already (in 2010) declared bankruptcy!

Software as a Service (SaaS) is one more example of an innovation based purely on a business model. Here, software packages are not sold to users or licensed for an annual fee, both of which involve a large capital or one-time expenditure and worries about maintenance and upgrades. Instead, the vendor provides the software package (with the responsibility of keeping it up to date and maintaining it), while the consumer pays only when it is used: a pay-per-use model, which converts software into a utility service, like water or electricity. In a somewhat similar manner, cloud-computing technology has led to the evolution of infrastructure as a service. Through this, users can avail of computing and data-storage facilities—again on a pay-per-use basis—without themselves making the capital investments that would have been required to buy the necessary equipment and software, or the expense and effort of maintaining them. Obviously, this is of tremendous benefit to smaller companies, which cannot afford the investments in information technology, nor do they have the in-house expertise to maintain it. Thus, the business model built around cloud technology has opened up vast new market segments for the IT industry.

These business models make services affordable by converting capital expenditures to operational expenditures, or work on a pay-per-use basis. Another innovation is a business model based on making greater use of resources that are not fully utilized by providing a platform that links users with providers. Technology is a big enabler in this new 'share' economy. One example is Uber, which has built its business around its own smartphone application.

This app links commuters to drivers willing to hire out their cars, enabling the former to ask for a hired car. The app shows the location of all nearby cars-for-hire, and the customer gets an immediate response about which one (with the name and phone number of the driver, and the car number) is coming to pick him or her up. It also indicates the likely time in which the car will arrive and enables the customer to track the car on his or her mobile. The hiring and payment is handled by Uber, with the fare being automatically billed to the customer's credit card at the end of the trip. In India, recognizing the uniqueness of the context, Uber allows payments to be made electronically or in cash (to the driver).

During high demand times, Uber increases its prices to attract more drivers, thus ramping up supply to meet demand. This is sound economics, but the surge-pricing model (sometimes seen as gouging) has come in for much criticism. This surge pricing is based on an automated algorithm, which responds immediately to changes in market supply and demand. While customers are informed about increased prices while making a reservation, it has nonetheless become an issue, especially when the fare has sometimes gone up to six or seven times the normal fare. In India, states are in the process of putting in place regulations that will cap such surge pricing. Partly in response, press reports[2] indicate that Uber will launch a feature that will inform users in India of the fare before booking a ride, replacing lightning bolts and pop-ups associated with surge pricing. This so-called 'upfront fare' will be first rolled out in six cities. While much can be said on both sides of this issue, it is one more example of technology outpacing the law, of innovation creating dilemmas for lawmakers.

As in most disruptive innovations, here too the incumbents (existing/conventional taxi services) oppose the new: in cities around the world (Indian cities included), traditional taxi-service providers are up in arms against Uber and its like. They demand that the same rules, regulations and licensing requirements that apply to them should be mandatory for the new app-based services. Obviously, there are no easy answers to this dilemma of new services that seem to fall foul of existing laws.

Uber was set up in 2009 by Garrett Camp and Travis

---

2. 'After Surge Pricing Row, Uber Moves to Make Fares Transparent', *The Hindu Business Line*, 25 June 2016.

Kalanick.[3] Officially launched in San Francisco in 2011,[4] the company soon added on operations, in the same year (2011), in New York, Chicago, Washington DC and—in its first foreign foray—Paris. Uber launched in India in 2013. By May 2016, it was available in sixty-six countries[5] and 483 cities.[6]

Uber is looking at future possibilities and, in February 2015, announced a collaboration with Carnegie Mellon University in the US to establish the Uber Advanced Technologies Centre to support the development of self-driving cars. This has borne fruit now, with Uber's launch of its driverless car service in Pittsburgh, USA, in September 2016. It has started with four cars, with each having two technicians, one of whom can take over driving if required. Uber's rollout of its innovation was preempted—by a few days—by a Singapore start-up (nuTonomy) which launched with six 'autonomous' cars in Singapore at the end of August 2016.[7]

Uber's other innovations include the launch, on a pilot basis in 2014, of Uber Fresh, a food delivery service, in the US, and a courier service called Uber Rush. Uber also introduced a carpool or shared service in 2014.

Despite the regulatory challenges it faces in many countries, Uber has continued to grow, not only in terms of customers, but also in recognition through value. Its innovativeness is

---

3. From the official Uber website.

4. From the official Uber website.

5. 'Uber slashes fares in 10 non-metro cities by up to 22%', *The Economic Times*, 12 April 2016.

6. From the official Uber website.

7. The 'driverless' launches by Uber and nuTonomy were covered in various media reports, including *The Times of India*, 15 September 2016.

acknowledged by investors who valued it at US$ 40 billion (in December 2014).[8] In a round of fundraising launched in 2016, the implied valuation shot up to US$ 66 billion.[9]

Like Uber, Airbnb too seeks to capitalize on unutilized resources by connecting these with potential users. It is an online marketplace that connects owners willing to rent out their property with users looking for space. Owners and users are categorized as 'hosts' and 'guests', both of whom must register—through a variety of means—with Airbnb. A valid email address and valid telephone number were initially the only requirements to build a unique user profile on the website. However, as of April 2013, a scan of a government-issued ID is now required.

Founded in 2008 in San Francisco, it began as Airbedandbreakfast (shortened in 2009 to its present name, Airbnb). The idea was apparently born when Brian Chesky and Joe Gebbia converted their living room into a bed-and-breakfast (with airbeds thrown in) and rented it out to delegates attending a conference. Later, Nathan Blecharczyk joined them as co-founder.

Beginning with a small investment from the well-known incubator, Y-Combinator, Airbnb has seen strong investor interest and has succeeded in raising funds at ever-higher valuations. Its growth has been as striking: from reaching 1 million bookings in 2011,[10] it crossed the 5 million nights

---

8. 'Making Sense of Uber's $40 Billion Valuation', *Harvard Business Review*, 10 December 2014.

9. 'Uber's books still top secret, but its biggest weakness isn't', *CNBC*, 8 June 2016.

10. From the official Airbnb website.

booking mark within a year thereafter and announced its 10 million guest nights booking in 2012.[11]

Airbnb went international in 2011, and by 2016 it had nineteen offices worldwide (two in the US), including New Delhi.

To help potential users, the Airbnb platform includes recommendations, reviews and ratings of properties by guests. In reverse, it has the hosts' reviews about their guests. The system of reviews by both parties allows guests and hosts to leave references and ratings, which are displayed to the public in order to provide an evaluation method. This has helped to build website credibility and serves as a useful input to future users. The site also provides a private messaging system as a channel for users to message one another privately before booking and accepting reservations.

Airbnb runs on a marketplace platform model where it connects hosts and travellers and enables transactions without owning any rooms itself. Thus, unlike traditional hotels (which own the rooms), Airbnb scales not by investing in creating more rooms, but by increasing the number of hosts and travellers, who are then linked with each other through its platform. Such platforms disrupt the traditional business model, and pose a new challenge to incumbent companies.

Recognizing the growing availability and use of mobile phones, Airbnb offers mobile applications for iOS and Android customers, as an alternative to its website. The mobile application downloads crossed the 1 million mark as early as September 2012, and accounted for over 26 per cent of the

---

11. From the official Airbnb website.

company's overall traffic.[12]

The mobile app offers users all the functionality of the website. This includes private messaging, which makes communication much faster between users. The application also allows users to find listings based on what is available, using geolocation.

In a further innovation, in June 2012, Airbnb launched a 'Wish List' feature, offering users the ability to create a curated catalogue of desired listings they would like to visit. This serves as a record of 'dream destinations' for the person, and can be shared with other users. Within a few months of the launch of the Wish List functionality in June 2012, the number of repeat daily Wish Listers increased by 3.5 times. About 45 per cent of users engaged with Wish Lists each day in 2012 and had added over one million accommodations to personalized lists.[13]

Given the disruptive and distinct business model, Airbnb—like Uber—faces regulatory challenges, including taxation and legal issues. Yet, it is perceived by the market as a business with great prospects. This is evidenced by its valuations. In April 2014, the company closed on an investment of US$ 450 million by TPG Capital. Thus, within five years, the privately owned company (founded in August 2008) had a valuation of approximately US$ 10 billion. In 2015, it raised US$ 1.5 billion, taking its valuation up to a whopping US$

---

12. 'Airbnb Mobile Usage Soars As Its iOS App Passes 1 Million Downloads, Accounts For 26% Of All Traffic', *Techcrunch*, 27 September 2012.

13. 'Airbnb users are enjoying Wish Lists, one million listings have been added to them', Drew Olanoff, *The Next Web*, December 2012.

25.5 billion.[14]

Uber and Airbnb exemplify new and disruptive business models based on innovation. Both use the capability of modern technology and its rapid spread. There are, though, other innovative business models that do not necessarily depend upon technology.

Many decades ago, long before the arrival of online marketplaces, a business model of a different nature was innovated. Conventional logic says that business is based on a producer selling a product or service to a buyer at a profit. The producer invests in plant, equipment, human resources and raw material; the buyer then pays a price (determined, ideally, by the market) to purchase the product or service. What if the consumer does not have to pay? This, most will instantly say, is unviable; only governments do this, as a means of providing goods and services (e.g., food, healthcare, education) to groups identified as needy or deserving. And of course, the government is not in 'business'. Now, think of commercial free-to-air television broadcasting: the audience (consumer) pays nothing to view the programmes, but the broadcaster makes a profit and is financially viable. A more recent example is a search engine like Google, which provides exceptional search capabilities at no cost to the user. In both cases, the secret of viability is advertising.

The advertiser, with a need to reach potential consumers, is willing to pay the TV broadcaster for carrying advertisements to the consumer. Enough advertising revenue ensures profitability for the broadcaster; the same for online search

---

14. 'Airbnb confirms $1.5 billion funding round, now valued at $25.5 billion', *Wired*, December 2015.

companies and radio broadcasters. This is an outstanding example of a business model that provides immense benefits at no cost to the user.

The use of commercials (advertising) as a means of funding programme content began many decades ago. The programmes, often referred to as soap operas or 'soaps', are generally serial dramas on television or radio, which feature multiple related storylines dealing with the lives of multiple characters. The stories in these series typically focus on emotional relationships to the point of melodrama. The name 'soap opera' stems from the fact that many of the sponsors and producers of the original dramatic serials' broadcast in the US on radio were soap manufacturers, such as Dial Corporation, Procter & Gamble (P&G), Colgate-Palmolive and Lever Brothers. The first serial considered to be a soap opera was *Painted Dreams*, which debuted on 20 October 1930 on Chicago's WGN. The first national radio soap opera was *Clara, Lu and Em*, which aired on the NBC Blue Network in January 1931.

In this model, the broadcaster or content provider basically provides a platform linking advertisers to consumers. This creates its own dynamics: TV programmes, for example, end up with 'lowest common denominator' content. It is to counter this, and to treat the audience as citizens rather than consumers, that public service broadcasting is important. Yet, the basic concept of a financially viable model in which the user does not pay is pregnant with possibilities. For example, it would indeed be a revolutionary breakthrough if someone were to innovate a similar model for social services and essential goods for the disadvantaged; a business model through which the provider made enough money to be viable, but the user paid nothing.

One interesting model in which the user pays nothing for

an important service is a conceived-in-India 'invention'. This is the unique and now ubiquitous 'missed call'. It is a simple way of conveying a pre-decided message, without the cost of texting (SMS) or calling, through a mobile phone. Thus, when you need to tell the driver that your meeting is over and could he please bring the car to the portico, you don't speak with him over the phone; instead, you call him up and disconnect before he answers. In any case, he will not take the call since he knows that the 'missed call' is the pre-designated code for him to bring the car to pick you up. Similar prearranged codes are used to signify events like 'I have arrived/reached'. A very effective, instantaneous form of communication, at no cost to the user. The service provider, though—unlike the television broadcaster—could not be too happy!

The 'missed call' innovation is indicative of the Indian penchant for finding ways of extracting the maximum value from any investment. In an example of how innovations build on each other, the missed call concept has been commercialized to create an innovative marketing model. Interestingly, this has been done not by an Indian, but by an American, Valerie Wagoner![15] The young woman, a California native, studied microfinance in graduate school at Stanford and became interested in emerging economies. She worked briefly with eBay's international marketing team and then moved to Bengaluru in 2008 to get closer to emerging market problems, 'which are more about need to have than nice to have', as she describes them. At mobile payments startup mChek, where

---

15. The ZipDial story is drawn largely from 'Why Twitter Bought Bangalore "Missed Call" Startup ZipDial', *Forbes Asia*, February 2015; and 'Twitter acquires Indian start-up ZipDial', *The Hindu*, 20 January 2015.

she worked with entrepreneur and venture capitalist Sanjay Swamy, she was intrigued by the missed call phenomenon. Brainstorming with Swamy and a third founder on monetizing the tactic, they created a company called ZipDial in 2010, based in Bengaluru.

By sending a text message to the originator of the 'missed call', the service connects product brands to their target Indian consumers. There is no charge for incoming text messages in India. This messaging mode is, therefore, especially useful for many in India (and other emerging markets) for whom the first online experience is increasingly through a mobile device, but the cost of data often constrains Internet usage. In less than five years (till 2015), Indians have missed-called ZipDial's numbers over 1 billion times to access promotions and information from brands like confectionary multinational Mondelēz's Cadbury's chocolates and Unilever's Dove soap.

ZipDial assigns each company a dedicated phone number, which the brands display in their ads or packaging. The customer can then give a missed call to that number and will get a text message (or even video content, on a smartphone). Disney India, for instance, has 2 million ZipDial missed callers who preview Disney content thirteen times per month on average. By comparison, on Twitter, Disney India currently has 30.1K followers. The advantage of using ZipDial is obvious.

When the horrific New Delhi rape case erupted in December 2012, Gillette launched a massive missed call crusade via ZipDial, asking men to pledge respect for women. Gillette ran the pledge in newspaper ads along with the ZipDial number to give a 'missed call' if they took the pledge, after which Gillette responded with another ZipDial number to refer friends to take the same pledge.

The early days were difficult for ZipDial, especially for Wagoner as the first foreign female CEO of an Indian tech start-up. ZipDial tasted success when it provided cricket scores and game content to millions of cricket-crazy missed callers. This helped net the first customer, Procter & Gamble, followed by a few others, and then venture capital began to flow in.

Growth was explosive and by 2015 ZipDial's roster exceeded 500 clients. It engaged nearly 60 million users through its platform with hundreds of marketer clients, including leading brands and media companies such as Procter & Gamble, Cadbury, Unilever, Colgate, Disney, MakeMyTrip and Amazon. Prominent Indian e-commerce players like Amazon, Flipkart and Uber have used the platform to push their mobile phone app downloads. By deducing the type of phone missed callers have, ZipDial is able to text them a link that takes them straight to the appropriate download page (which cuts down on costs, since the consumer doesn't have to surf to find the relevant page).

Banks have gotten into the act (say, to check your balances). Politicians like Prime Minister Narendra Modi, Bollywood stars Amitabh Bachchan and Shah Rukh Khan and cricket icon Sachin Tendulkar have seen massive surges in their digital following as millions of their fans give a 'missed call' to receive their idols' tweets, audio messages and photos on their phones.

In January 2015, marking its first acquisition in India, Twitter announced the acquisition of ZipDial. While the size of the deal was not disclosed, reports peg it at about US$ 30 million.

Apart from extracting maximum value from any expenditure ('paisa vasool'), another great Indian game—one played with especial vigour and relish—is to find methods by which to beat the system. In most instances, this is driven by need rather than by any deep rebelliousness: given the over-abundance of laws and regulatory restrictions, many of which do not take into account the context or genuine requirements, life in India would be difficult for the common man without these system-beating bypasses. Sometimes, these innovations exploit flaws and distortions in the system, or the inevitably large gaps between supply and demand, between urgency of need and systemic delays in meeting it.

Some decades ago, I came across a rather strange story of one way to beat the system. To help understand it, it is necessary to explain the context. 'Locals'—as the commuter trains in Mumbai are called—serve as the lifeline of the city, ferrying millions of passengers every day, mainly between home and workplace. Though an extremely efficient service, the massive volume of traffic leads to trains being so crowded that passengers are packed tighter than sardines in a can. Some trains are designated as 'fast' and a few as 'super-fast'—these are express trains that skip some or many stations, thereby reaching their destination much quicker. A majority of the commuters are regular travellers who have a 'pass' (local parlance for a monthly or season ticket), but there are also huge numbers of occasional passengers who need to buy a ticket for their journey. The veteran commuter is able to time her arrival at the station just in time to board the regular 8.13 a.m. 'super-fast' to Churchgate (the train terminus in South Mumbai). The occasional traveller, wanting to catch the same train for a vital appointment in South Mumbai, arrives at the station early enough, but sees a long queue at the ticket window. After a few impatient minutes in the queue, he realizes that he will not be able to buy a ticket in time to catch the 8.13. The next two trains are 'slow' ones (stopping at all stations), and will never get him to Churchgate in time for his meeting. Desperate, he decides to give his conscience a brief rest and boards the train without a ticket. He is, of course, aware that the chances of being caught by the travelling ticket-checker are small, given the crowd in the train. The hordes that disembark at Churchgate (a few thousand from each train, with one arriving every two to three minutes) almost certainly ensure free passage there too. However, if caught, the penalty is hefty, many times the cost of the ticket. There are always a fair number of people who, like the traveller in this case, travel without a ticket, due to force of circumstance. However, there are also many who do so routinely, gambling on the chance that they will not get caught, yet fearful of the huge cost if they do.

It is in this scenario that a clever entrepreneur created an innovative business based on the principle of insurance. Without hiring any top-notch actuarial talent, he worked out a monthly insurance premium to be paid by a person desirous of being insured against the chance of getting caught and fined for ticketless travel. The entrepreneur would reimburse the fine paid by any of his customers who was nabbed by the ticket inspectors. He soon built up a roaring and profitable business, with a rapid growth of clientele who did not want to spend time or money on buying a ticket. Many daily commuters too swapped their season ticket for an insurance policy. Apparently, it was only a tough crackdown on the entrepreneur combined with very intensive ticket-checking (raising his costs) that ended this illegal—though doubtless innovative!—business model.

This story has an interesting sidelight. On the one hand, the business model encouraged ticketless travel and illegal behaviour; on the other hand, it brought—even if in a somewhat distorted manner—moral action. For every 'insured' ticketless traveller who got caught had to produce proof of having paid a fine in order get it reimbursed by the insurer. Earlier, in most cases, ticketless travellers who were accosted bribed the inspector. Thus, instead of paying the specified fine of Rs 500, a smaller amount of Rs 100 or Rs 200 would change hands. The ticket inspector was happy to make some money and the traveller was glad to get away by paying a smaller amount. However, there was obviously no proof of having paid, nor of the amount involved. The insurance scheme, though, needed proof of payment: the traveller had to produce an official receipt to claim reimbursement whenever he was caught. Therefore, when any of those insured were apprehended for travelling without a ticket, they would pay the fine, but insist on a receipt. This meant that the money could not be pocketed by the inspector, though it also required payment of the full official fine amount. The penalty, instead of adding to the inspector's personal income, now went to the railways. So, through a rather convoluted way— and an illegal scheme—corruption was reduced!

While the Mumbai local train story is a few decades old, it was surprising to read a report[16] of a similar model from current times—this one from Sweden; rather unexpected from a country reputed for its honest citizenry and a social security system that provides little reason to cheat. An organization in Stockholm, named Planka.nu (meaning 'free ride now') recruits members who are required to fulfil two conditions: pay a monthly fee of about US$ 15, and promise to evade payment

16. 'Fare Dodging Is an Organized Rebellion in Stockholm, and It's Winning', *The New York Times*, 17 May 2014.

every time they travel in the Stockholm Metro. Like the Mumbai 'insurance' scheme, this group too reimburses any of the fines (approximately US$ 180) that may be due if a member is caught. The US$ 15 monthly fee is small compared to the ticket price of about US$ 120 for a thirty-day unlimited-ride pass (equivalent to the monthly pass in Mumbai). Apparently, Planku.nu offers instructional videos on how to slip through station gates without paying! Obviously, this will help reduce their payout. Stockholm Public Transport, which operates the metro system, estimates that as many as 15 million trips in 2013 were not paid for. Clearly, the business model of Planku. nu is working rather well!

These examples raise the question: are crooked minds more innovative? ('Crooked' here deriving from 'crook'.) While this is debatable, it seems clear that innovation requires crooked thinking ('crooked' here in the geometric sense)—i.e., thought processes that deviate from the straight and conventionally defined mode. After all, an innovation means something different from what exists: be it a product, a process or a business model. It takes what exists (a washing machine, for example) and puts it to a different use (possibly, to churn lassi). It takes a systemic problem and context to create a new business (insurance for ticketless travellers). It takes new technology (like cloud computing) and creates a new business model (pay per use). It creates a new business model (advertisers paying, not consumers) with existing systems and technologies. Innovation takes many forms, but all of them are based on going beyond the straight and narrow, on doing something differently, based on thinking differently. It creates value—generally, disproportionately large value—from what already exists or from creating something new. Innovation is therefore a value creator or value multiplier, often enhancing returns with little or no additional investment.

A disproportionately high return on investment—in the business or economic sense—is certainly one of the hallmarks of innovation. However, just as it is not merely about new technology or products, nor even about only creative business models or processes, innovation is not limited to economic returns. There are some outstanding examples from the artistic sphere that can be called either creative or innovative, depending on one's definition of these terms. Artists would, in all probability, prefer the term 'creative' and undoubtedly each of these is, in many ways, an exemplification of creativity. Yet, to the extent that they build on and borrow from what already exists, one could equally classify them, without in any way detracting from their impact, as innovations.

In the arena of writing, there has been much creative expression over many centuries. Outstanding authors have not only created and enhanced the richness of our cultural heritage, but more than a few have had an impact on our thinking. The innovation I would like to highlight, though, is not the *content* of the many great and memorable literary works, but in the *form* of writing. Vikram Seth used the complex poetic format of iambic tetrameter to write a whole novel, *The Golden Gate*.[17] For those with a scholastic bent of mind, 'tetrameter' is a line of poetic verse that consists of 4 metrical feet. In English versification,[18] the feet are usually iambs (an unstressed syllable followed by a stressed one, as in the word 'belcause'), trochees (a stressed syllable followed by an unstressed one, as

17. Vikram Seth, *The Golden Gate*, Random House, 12 March 1986.
18. The explanation that follows is from the *Encyclopedia Britannica*.

in the word ti′l ger), or a combination of the two iamb sounds like a heartbeat, sort of like 'duh-DUH'. When four beats are placed together in a line of poetry, it is called tetrameter. When we combine iamb with tetrameter, it is a line of poetry with four beats of one unstressed syllable followed by one stressed syllable, and it is called iambic tetrameter. It sounds like: duh-DUH, duh-DUH, duh-DUH, duh-DUH. Some believe that tetrameter is a natural rhythm and that it is easy to read out loud. The innovative format of *The Golden Gate* was inspired by Charles Johnston's translation of Alexander Pushkin's Russian classic, *Eugene Onegin*, written in 1833. Arguably, this unique form of presentation has enhanced the impact of Vikram Seth's novel.

As in writing, the visual medium too has seen a high order of creativity and work that has had a great impact. Anyone who has seen the picture of the little girl with flaming clothes, running from the horror of napalm bombing in Vietnam, will never forget the image, nor the terrible tragedy and human suffering inflicted by war. This is, in many ways, akin to the content, rather than form, of a book; representing, therefore, creativity rather than innovation. However, there are other visual representations that embody innovations too.

One example is the popular, widely used caricature of Mahatma Gandhi. It shows his profile with but a few lines, which outline the shape of his (hairless) head and his round wire-frame spectacles. Through innovative use of the not-uncommon device of 'negative space', this very simple artwork effectively creates a picture that is almost instantly, and near-universally, recognized as Gandhi. Certainly an artistic innovation of a high order!

A similar use of negative space and iconic symbols is the

caricature that uses a bowler hat, a small moustache and a walking stick to create the immediately recognizable image of Charlie Chaplin.

An outstanding example of innovation in using visuals to convey a deep message comes from the master filmmaker Satyajit Ray. In *Mahanagar* (1963), the daughter-in-law becomes the breadwinner of the joint family after her husband loses his job. The consequent change in the relationships within the family is highlighted through just one visual shot—as the young woman sits down for a meal, her mother-in-law serves her the head of the fish. The uninitiated may not even notice this small nuance, but to anyone familiar with the culture and traditions of Bengal, this one shot speaks louder and clearer than a thousand words. For, in Bengal, the fish-head is considered a delicacy and is normally served to the head of the household. Once one knows this, the significance of the mother-in-law's gesture hits you with hurricane force. What an innovative use of semiotics!

Another interesting out-of-the-box idea from artists—though not pertaining to art, per se—originates in Austria.[19] Sculptor Fritz Gall, along with four other local artists, co-founded the 'Nonseum' in 1994. The genesis was Austria's first fair of failed inventions, held in 1984. Its unexpectedly huge success made Gall and his friends realize that they had uncovered a unique market niche—coming up with seemingly pointless inventions, which manage to serve the purpose of making people pause and think.

The five artists organized—over a decade from 1984—a

---

19. This outline draws from 'Nonseum, a museum of the absurd with a cult following', *The Hindu* (Delhi), 16 August 2015.

series of apparently absurd events, like a twenty-four-hour snail race or a festival of scarecrows. As their inventory of weird inventions grew, they finally decided to unite all these silly ideas under one roof. Thus was born the Nonseum, with its vast collection of absurd-yet-brilliant inventions, which now attracts thousands of people from around the world to its buildings in the small village of Herrnbaumgarten, in the wine region near the Czech border.

Science and technology are well-recognized sources for innovation, as was briefly noted earlier in this chapter. The amazing ideas and products that have, over the centuries, emanated from scientific and technological effort embody innovation at its best. Many of these have had far-reaching impact, not only on the economy but also sometimes on society and social life. Apart from the output of science and technology, it is interesting to note that the method of science or the actual process of scientific research has itself seen some exciting innovation. The method of science may be blandly described as hypothesis–observation–measurement–experimentation–analysis–theory–testing–validation/modified hypothesis. Yet, even this seemingly fixed and straightforward process has required and witnessed innovation. Here is one recent example of such an innovative approach.[20]

It is well-known that exercise can make us fitter and reduce the risk of illnesses like diabetes and heart disease. But how exactly a run or bike ride might translate into a healthier life has remained a mystery. Exercise, a new study finds, changes the shape and functioning of our genes, which is important for

---

20. What follows draws substantially from an article published in *The New York Times*, 17 December 2014.

improved health and fitness. Scientists know that certain genes become active or quieter as a result of exercise. But for a long while they did not understand the process through which those genes knew how to respond to exercise. Decoding this requires the study of epigenetics, the process by which the operation of genes is changed, but not the DNA itself.

Epigenetic changes occur on the outside of the gene, mainly through a process called methylation. In methylation, clusters of atoms, called methyl groups, attach to the outside of a gene like microscopic mollusks and determine its ability to receive and respond to biochemical signals from the body.

Scientists know that methylation patterns change in response to lifestyle. For example, certain diets or pollutants can change methylation patterns on some of the genes in our DNA and affect what proteins those genes express. Depending on which genes are involved, it may also affect our health and risk for disease.

Since methylation patterns are affected by a whole host of lifestyle factors, it had been difficult to isolate the effect of exercise alone on epigenetic changes. Scientists from the Karolinska Institute in Stockholm overcame this problem (of isolating the effects of exercise from those of lifestyle—diet, behaviour, etc.) through an innovative methodology. Since the complex factors that determine lifestyle are difficult to control for amongst different individuals, the standard methodology of using one set of individuals as a control group is not adequate. To get over this difficulty, the researchers recruited twenty-three young and healthy men and women, and put them through a series of physical performance and medical tests, including a muscle biopsy, in the lab. They were then asked to exercise half of their lower bodies for three months, by bicycling using only one leg, leaving the other unexercised. In effect, each person became his or her own control group. Both legs would undergo methylation patterns influenced by his or her overall lifestyle; but only the exercised leg (the one used for pedaling) would show changes related to exercise.

This experiment helped the researchers to determine, through sophisticated genomic analysis, that more than 5,000 sites on the genome of muscle cells from the exercised leg now featured new methylation patterns. Some showed more methyl groups; some fewer. But the changes were significant and not found in the unexercised leg.

As a result of this experiment (published in *Epigenetics* in December 2014), scientists now have a deeper understanding of one more step in the complicated, multifaceted processes that make exercise so good for us. This has come about thanks to an innovative idea—simple, but impactful like many truly innovative approaches—of using parts of the same individual as elements of the control and experimental group.

Such creative thinking is, fortunately, not limited to scientists, writers, artists or sharp businessmen. Policy makers and city administrators too can sometimes come up with out-of-the-box ideas. Here are two such examples.

The first is a rather creative solution in the social sphere from Columbia.[21] Faced with the common situation bedevilling many urban areas—massive problems and meagre budgets—the then mayor (from 1995 to 2003), mathematician Antanas Mockus, came up with a radical solution. He dissolved the deeply corrupt traffic police force and replaced it with 420

---

21. This example is from 'Dramatic Solutions', *The Economist*, 11 July 2015.

mimes who directed traffic and poked fun at jaywalkers. Drivers were provided with red and white cards to punish or praise other road users.

There is no real evidence that such tactics did much good, but they apparently did little harm. Bogota's murder rate dropped by 70 per cent between 1995 and 2003, and traffic deaths fell from 1,300 a year to 600. This may well have been due to other extraneous factors. Unconnected or false correlations are not uncommon, and no one knows this better than mathematicians—like Mayor Mockus. A more controlled experiment, or a longer period, may well have helped to prove (or disprove) the value of this innovation. Unfortunately, though, there is no way of knowing for sure, as Mr Mockus's idea of fighting loutishness with theatre fell out of favour when he left office. But more recently Gustavo Petro, Bogota's left-wing mayor, called for an encore: in January 2015, Bogota's transport office once again hired three troupes of actors to fan out across the city daily to educate, cajole and in some cases shame anti-social Bogotanos. People who barged on to buses before others got off might have been met with intimidating stares from actors planted among the passengers. In one skit, played out on an articulated TransMilenio bus, an actor on a mobile phone described how his cousin was hit by a bus as he tried to dash on to another without paying.

The revival of street theatre may well have only a limited run. Critics say it is a waste of money. Bogota's next mayor may bring down the curtain. Yet, this rather intriguing approach represents a very different and innovative initiative, the kind of creative thinking that is particularly necessary to tackle the many intractable social problems facing all countries.

The second example of creative city administrators is

from India. Many years ago, a small innovation—so obvious, in retrospect, that one wonders why it is not adopted more widely or wasn't done earlier—helped to hugely decrease traffic congestion in Mumbai. At that time, much of the commercial activity of the city—offices, shopping areas, major hotels—as also the university was concentrated in the southern end of the city. This meant that much of the traffic would flow from the north (the 'dormitory' suburbs) to the south each morning, and in the reverse direction in the evening. Being a narrow island, a large part of the traffic was carried by one main arterial road. The simple innovation was the change from the standard equal division of the road to an unequal one; each morning, during peak hours, an additional lane was provided for the traffic going south by reducing the road availability for traffic in the reverse direction; in the evening, road space was divided in the opposite manner. This optimization of a scarce resource (road space) increased the carrying capacity in the desired direction by over a third, at practically no additional cost. All it required was the temporary placement of bollards for a few hours each day, to mark out the additional lane for the traffic.

Urban roads provide another, though negative, example of the Indian genius for innovation. This is based on attempted optimization at the level of the individual and is on view when traffic is very heavy. A traffic jam in one direction immediately leads to cars being driven on the wrong side of the road, emulating illegally, and therefore with disastrous consequences, the positive innovation mentioned earlier. Motorcycles often use the pedestrian footpath to get ahead of slow-moving traffic on the road. Vehicles waiting at a railway crossing will inevitably occupy both sides of the road, leading to head-on 'confrontations' and a chaotic jam the minute the crossing opens. At a failed traffic signal, a gridlock takes place in moments as each vehicle, with no heed for traffic rules, leave alone courtesy, follows a me-first philosophy. More self-discipline and orderly traffic would increase the average speed and reduce accidents, benefiting all. These are good examples of how individual 'innovation' can be disruptive, dysfunctional and sometimes socially regressive, providing a contrast to the beneficial-to-all 'extra-lane' initiative mentioned earlier. This brings home the challenge of how to channelize individual out-of-the-box thinking for broader organizational or societal good. What structures and rules can best help this? Are there policies that can ensure this and guide or direct innovative energy into desirable directions? Is there a framework for a facilitative ecosystem to promote innovations that are beneficial to the organization? These questions and related issues are addressed in later chapters.

3

IDEAS TO BENEFITS

Gains from brains. Money, networking and mentoring are crucial, but so is the ecosystem. Examples of efforts in India to spur start-ups and innovation.

Every society, every country, has individuals who are innovative. One challenge is how best to translate such innovativeness into products or services that can be of social or economic benefit. A deeper and longer-run challenge is to identify ways in which the number of innovators can be increased. The holy grail is the creation of an innovative society.

These broad challenges can be broken up into a number of specific elements. Immediate returns can be realized by capitalizing on low-hanging fruit—identifying innovations that can be quickly translated into products or services and providing the necessary support to do so. At first glance, there always seem to be many innovations that only need to be put into production and distributed. Inevitably, though, one finds that most are not quite ready for the market. Often, one has to tweak them in order to get to mass production, or features need to be added to make a product more easily accessible to the consumer. The business model, financing and marketing have to be thought through and sales, distribution and maintenance issues must be resolved. The difficulties involved are exemplified by the small number of innovations that successfully make the journey from laboratory to market.

One example is the electric rickshaw. For many years, various innovators have been working on alternatives to the conventional cycle rickshaw and autorickshaw, looking for a non-polluting, inexpensive, efficient and fast means of transportation. Many products were developed, but for one reason or another, none of them were able to create a market for themselves. The gap between a laboratory model and market success needed to be bridged by addressing issues related to design, finish, cost and easy production. A laboratory model or a proof-of-concept is not, by itself, one that can be produced, marketed and actually bought by a user. It is often this stage—from a working prototype to a marketable product that has a demand—that is the most difficult one. In many cases, this stage requires as much innovation as does the creation of the product itself.

In the case of the electric (battery-powered) rickshaw, it was only after many years of research that the technologists of the CSIR came up with a design that addressed many of the problems. Soleckshaw (solar-electric rickshaw), unveiled in October 2008, is an eco-friendly tricycle[1]. It is driven partly by pedal and partly by electric power supplied by a battery that is charged using solar energy. Soleckshaw is the brainchild of Prof. Samir K. Brahmachari, director-general of CSIR from

---

1. 'The humble cycle rickshaw gets a solar-powered makeover', *Mint*, 14 October 2008.

2006 to 2012.[2]

The Mark I version of Soleckshaw has been designed, developed and prototyped by one of CSIR's national laboratories, the Central Mechanical Engineering Research Institute (CMERI) at Durgapur. The accompanying solar-charging station for the batteries has been set up by Central Electronics Limited (CEL), a Government of India undertaking. The specifically designed robust, low-power, high-torque, brushless DC motor has been developed for the first time in India by Crompton Greaves, based on the specifications provided by CMERI.

The Advanced Materials & Process Research Institute (AMPRI), another constituent laboratory of CSIR, provided support for the Technology Demonstration Project (TDP) at Chandni Chowk in Delhi. AMPRI's fly ash jute-polymer composite is used to house the batteries. The Centre for Rural Development (CRD), an NGO working on rickshaw banks in different cities, has joined the TDP as a partner for wider deployment of Soleckshaw. The Delhi Metro Rail Corporation (DMRC) has provided the facility for accommodating the solar charging unit at its Metro station at Chandni Chowk. The Municipal Corporation of Delhi (MCD) has further facilitated the TDP by allowing necessary construction in Yudhbir Singh Park and its use for a period of one year. The synergy among the above organizations reflects the importance and benefits of teamwork, and its importance in taking forward an innovation.

Robust and ergonomically designed to take the drudgery

---

2. Some of what follows is based on a press release by the Ministry of Science and Technology, '"Solekshaw" Eco-Friendly Dual-Powered Rickshaw Launched', 2 October 2008, and conversations and inputs from Prof. Samir K. Brahmachari.

out of rickshaw-driving, the prototypes of Soleckshaw were flagged off at CMERI, Durgapur, by Prof. Brahmachari on 17 August 2008.

With better aesthetics and ergonomics, the cost-effectiveness of Soleckshaw were engineered by optimizing the system around the most appropriate commercially available components. This would also minimize the capital requirement for a mass manufacturing unit. Only the novel sub-assemblies like the differential drive, the special hub motor with regenerative feature and the lightweight solar panel need to be specially manufactured, apart from the chasis designed for a comfortable ride even for senior citizens and the physically challenged. Ensuring financial viability of the Soleckshaw will need an innovative business model. This is being evolved with the involvement of NGOs, banks, environment-friendly corporates and manufacturing organizations. The aim is to bring the cost down to the level of an ordinary cycle rickshaw. Although CSIR held the intellectual property rights (IPR) on the design of the Soleckshaw, it adopted a crowd-sourcing effort by marketing it under a non-exclusive licence, through which the licencees were provided the IPR at a nominal charge and allowed to modify the original design. The consequence of such a policy decision and the crowd-sourcing effort was the rapid proliferation of the electric rickshaws in the market.[3]

The real comparison—and competition—is between an auto-rickshaw and an e-rickshaw. An e-rickshaw costs around Rs 85,000, while an auto-rickshaw is priced at Rs 1.68 lakh. The e-rickshaws operate on four batteries, which last for six

---

3. Narayanan Suresh and V. Dixit, 'Solar Powered Rickshaws', *MIT Technology Review*, Massachusetts, 2009.

months. A new set of batteries is available for Rs 25,000. These batteries need to be charged overnight and can operate for 80 km. An autorickshaw, on the other hand, requires Rs 100 worth of fuel (compressed natural gas) to run for a whole day. As a result, e-rickshaws charge Rs 10 for a distance between 2 and 5 km, while auto-rickshaws charge Rs 25 for first 2 km, and Rs 8 for every additional km. E-rickshaws should be driven at a speed of 25 kmph, but most of these vehicles travel at speeds between 20 and 35 kmph. Autorickshaws, of course, can go much faster and can travel at a maximum speed of 60-70 kmph.[4] However, in city traffic, the actual speeds end up being almost the same.

CSIR ensured that they adhered to the 250-watts stipulation of the Motor Vehicles Act, 1988, so that the Soleckshaw could be categorized as a 'Non Motor' vehicle. As a result, Soleckshaw drivers do not require a licence or registration. Being a low-powered rickshaw, the Soleckshaw can carry two persons, or a maximum weight of 200 kg (excluding the driver). More recently, a new breed of electric rickshaws have begun plying. These are powered by motors of anywhere between 600 and 1,000 watts, thereby allowing a capacity of four persons. This has stirred a debate of whether or not they should be brought under the 'Motor Vehicle' category, requiring registration and a licence to ply the rickshaw. In such a scenario, the rickshaw drivers may not have the basic educational qualifications required (as stipulated in the Motor Vehicles Act) to obtain a licence. There is, however, pressure to relax this requirement.

---

4. From 'In Delhi, it's E-rickshaws vs Auto-rickshaws', *The Indian Express*, 23 June 2014.

By 2014, the proliferation of electric rickshaws led to a policy intervention, with the Government of India recognizing the large employment being created through the e-rickshaws and therefore deciding to amend the Motor Vehicles Act of 1988, bringing the e-rickshaws under the legal framework.[5] This will now require them to obtain a registration, but has the advantage of removing the legal ambiguity about their status (and the harassment this sometimes entailed).

This example highlights how technologies and products that seem to be low-hanging fruit—to stretch the metaphor—may not always be ripe; they need attention, care and treatment before they can be harvested. Therefore, from a large list of exciting and worthwhile innovations, one has to carefully identify those that are potentially marketable and then spend time and effort to take them to the stage of ensuring consumer demand.

In a large and poorly interconnected country like India, it is not an easy task to put together a database of such innovations. One organization that has, on a sustained basis, tried to do this, is the National Innovation Foundation (NIF).[6] This is an autonomous body under the Government of India's Department of Science and Technology. It was set up in February 2000 in Ahmedabad to provide institutional support for scouting, spawning, sustaining and scaling up grass-roots innovations. NIF is the brainchild of Dr Mashelkar (formerly director-general of CSIR) and Prof. Anil Gupta, professor at the Indian Institute of Management, Ahmedabad (IIM-A),

---

5. Nisha Chandran and S. K. Brahmachari, unpublished work, 2015.

6. What follows on NIF and related activities draws a great deal from inputs provided by Prof. Anil Gupta.

and the pioneer of locating and showcasing innovations at the grass-roots level.

NIF conducts a biennial national competition for grassroots green technologies developed by farmers, mechanics, artisans and others through their own genius, without any recourse to professional help. NIF validates these innovations with the help of experts, and ascertains their novelty by doing a search of 'prior art'. If the innovation is deemed novel, NIF files a patent on behalf of the innovator. NIF also funds value-addition initiatives for these innovations to upscale them and make them more useful for a larger segment of people.

To determine the feasibility of commercializing a technology, NIF conducts market research and test marketing. Those technologies that are found to be commercially viable are licensed to willing entrepreneurs. A Micro Venture Innovation Fund, sponsored by Small Industries Development Bank of India in 2003, supports prototype development, test marketing and pilot production.

IGNITE is an annual competition for students' ideas and innovations conducted by NIF in partnership with the Central Board of Secondary Education and some state education boards. All students up to the twelfth class from any school (and out-of-school youth of the same age group) in India are eligible to participate in IGNITE.

NIF is mandated to build a national register of ideas, innovations and traditional knowledge practices related to agriculture, plants, animal health and human health. With the help of the Honey Bee Network, NIF has been able to scout and document over 181,000 examples of technological ideas, innovations and traditional practices. Since its inception, NIF has also recognized over 550 grass-roots innovators and students in its various national award functions, providing them a platform to showcase their creativity.

The core principles of NIF stem from the Honey Bee Network and the Society for Research and Initiatives for Sustainable Technologies and Institutions (SRISTI). The Honey Bee Network is a volunteer network spread across seventy-five countries, which is engaged in developing a sustainable knowledge ecosystem. SRISTI, a non-governmental organization, was set up in 1993 to help document the innovations and traditional knowledge practices discovered by the Honey Bee Network. The Honey Bee Network and SRISTI were both founded by Prof. Anil Gupta.

Another initiative of Prof. Gupta is the Shodh Yatra (journey of discovery, or a pilgrimage of search), of which thirty-five have been undertaken in the last fifteen years. Each of these journeys or trips involves a group of students, led by Prof. Gupta, trekking to various—often remote—parts of India to look for and then recognize, respect and reward creative and innovative people and communities at their doorstep. The group also shares its database on sustainable technologies, even as it learns about local innovations and challenges. The idea is to share first and then seek; to cross-pollinate ideas and minds, in keeping with the philosophy of the Honey Bee Network.

The Shodh Yatra has evolved into a regular course, too, at IIM-A (where Prof. Gupta teaches). Recognizing its innovative nature, the course was approved, though it did not have a session-by-session plan and does not permit any course-reading material or book to be taken along for the journey. The philosophy is that it entails learning from four teachers—from within, from peers, from nature and from common people. The course consists of a journey to the Himalayas, and involves reflective learning about leadership for social change. It has been very successful and is very popular amongst the students.

Shodh Yatra and the course are not merely ways of discovering grass-roots innovation and learning from it (hopefully, being inspired to be innovative), but themselves reflect a new and innovative approach.

In looking at innovation at the grass-roots as well as other levels, it is clear that funding is amongst the essential elements of taking an innovation from prototype to marketable product. In areas and countries where the innovation ecosystem is well-developed, such funding is available fairly easily. Early-stage venture funds exist in fair number, and there are also so-called 'angel investors'—individuals and organizations—who provide funding at the initial stages of an innovation. At this point, the success of the innovation is uncertain, and the funder is taking a chance based on an assessment of how well the product might do in the marketplace. This risk is high and there is a fair probability that the investment will turn out to be completely infructuous. It is for this reason that, in some countries, there are tax incentives available for such funding.

In India, though the situation is rapidly improving, funding for the initial stages is yet scarce. Part of the reason is the lack of tax incentives. An additional factor is the lack of easy exits for such early stage funders. This is important because investments in new and innovative ideas are not only very risky, but have a long gestation period. There must, therefore, be clear exit pathways for early investors who want to divest part (or all) of their investment in a year or two.

This requires the ecosystem to have a well-developed exit mechanism through which early investors can sell their stake. In fact, there is a strong argument for focusing on creating easy exit options rather than trying to promote angel or early-stage funding: the former may be easier to do and will automatically create enough pull effect to attract angel investors at the start-up stage. Easy exits, apart from tax incentives, are often cited as one reason for the robust availability of funding for start-ups in the US.

Funds are, obviously, a critical factor for any enterprise or entrepreneur. However, a start-up requires much more than an idea and money. An important element, especially for young or first-time entrepreneurs, is mentorship. Guidance from experienced experts, handholding through critical phases, providing a sounding board, and advice and inputs from knowledgeable critics are amongst the vital inputs from mentorship that often make the difference between success and failure.

Networking, either through a mentor or separately, is one more facet that is of great value. The connections that must be established with suppliers and customers are obvious, but equally important are the linkages that need to be established with complementary enterprises, or even with competitors. Today, in many areas, success is not necessarily determined by the quality, cost or desirability of a product; it is, instead, the outcome of other environmental factors. A striking example of this is the field of mobile telephony. Some companies concentrated on creating great devices; their mobile telephones were well-designed, had many technological innovations and were available at different price points. Initially, they were successful, but after a while, disaster struck. They had focused on their product, forgetting that it was but one element of an overall system; a change in the system rendered their product vulnerable, even obsolete. Nokia and Blackberry are amongst the victims. Both were very successful companies, with their products being amongst the market leaders, but they overlooked or ignored the change in the ecosystem within which mobile devices worked.

Indicative of their rise and fall are their stock prices. Blackberry (Research in Motion) was trading at US$ 149.90 per share on 19 June 2008, but five years later (9 December 2013) it was down to as low as US$ 5.75.

Similarly, Nokia nose-dived from a peak of US$ 230 (in March 2000) to a low of US$ 1.69 in July 2012. In response to its growing problems, Nokia announced in September 2010 that Stephen Elop would take over as CEO. His reading of Nokia's situation, the missed opportunities and strategic challenges, was articulated in a much-quoted letter to employees (generally referred to as the 'burning platform' memo) sent out a few months later. He recounted a man's dilemma:[7]

> There is a pertinent story about a man who was working on an oil platform in the North Sea. He woke up one night from a loud explosion, which suddenly set his entire oil platform on fire. In mere moments, he was surrounded by flames. Through the smoke and heat, he barely made his way out of the chaos to the platform's edge. When he looked down over the edge, all he could see were the dark, cold, foreboding Atlantic waters.
>
> As the fire approached him, the man had mere seconds to react. He could stand on the platform, and inevitably be consumed by the burning flames. Or, he could plunge 30 metres into the freezing waters. The man was standing upon a 'burning platform', and he needed to make a choice.
>
> He decided to jump. It was unexpected. In ordinary circumstances, the man would never consider plunging into

7. The quotes from Elop's memo are from 'Nokia's chief executive to staff: "we are standing on a burning platform"', Charles Arthur, *The Guardian*, 9 February 2011.

icy waters. But these were not ordinary times—his platform was on fire. The man survived the fall and the waters. After he was rescued, he noted that a 'burning platform' caused a radical change in his behaviour.

We too, are standing on a 'burning platform,' and we must decide how we are going to change our behaviour.

Elop pointed to how competitors had quickly moved ahead. For example, Apple's market share in the US\$ 300+ price range had gone up from 25 per cent to 61 per cent in just two years (from 2008). Android, he noted, had

> ...created a platform that attracts application developers, service providers and hardware manufacturers. Android came in at the high-end, they are now winning the mid-range, and quickly they are going downstream to phones under €100. Google has become a gravitational force, drawing much of the industry's innovation to its core.

At the low end, manufacturers in China, making phones at unbelievable speed and cost, accounted for more than a third of the phones sold globally. Highlighting the importance of pace of innovation, Elop said, 'We have some brilliant sources of innovation inside Nokia, but we are not bringing it to market fast enough.' Later in the memo he reiterated, 'We haven't been delivering innovation fast enough.'

Finally, in an emphasis on the ecosystem, he said, 'For example, there is intense heat coming from our competitors, more rapidly than we ever expected. Apple disrupted the market by redefining the smartphone and attracting developers to a closed, but very powerful ecosystem.' He also said,

> The battle of devices has now become a war of ecosystems, where ecosystems include not only the hardware and software

of the device, but developers, applications, e-commerce, advertising, search, social applications, location-based services, unified communications and many other things. Our competitors aren't taking our market share with devices; they are taking our market share with an entire ecosystem. This means we're going to have to decide how we either build, catalyse or join an ecosystem.

While Elop was specifically addressing issues related to Nokia, two factors are of importance for innovation in general. The first is the speed or pace of innovation. It is not enough to be innovative; it is also necessary to innovate at a rapid clip. In today's rapidly changing world, such speed is critical. The second factor is the ecosystem. This can spur innovation or derail it. Innovations can change the ecosystem itself. On the other hand, an innovation that does not fit into an emerging ecosystem will just not survive.

The existence of a conducive ecosystem is central to innovation. It is the soil in which the seed of the idea will either take root or perish. From around the start of this century, many governments as well as organizations within various countries have attempted to create supportive ecosystems. In India, angel investors and a small but growing number of incubators and accelerators have worked on creating the appropriate support conditions for companies in which they invest. Apart from funding, they provide mentorship and facilitate networking. Incubators and accelerators also provide physical facilities, with the physical proximity of other start-ups serving to promote a peer-level exchange of experience and ideas. Organizations like NASSCOM, TiE, iSPIRT and others have also been key players in the entrepreneurship-innovation ecosystem.

Amongst these is the Indian Angel Network (IAN),[8] India's first angel investor group, set up in April 2006, which brings together successful entrepreneurs and CEOs who share a passion to enable early-stage businesses to create scale and value. It not only provides money but also mentoring and inputs on strategy and execution. IAN members also provide investee companies access to their own business networks, to help the companies scale and grow.

IAN has now become the world's largest angel group, in terms of members, number of investments and amount of investments. It has around four hundred members, drawn from across the country and some from overseas, including leading lights from diverse sectors, comprising the who's who of successful entrepreneurs and CEOs. IAN was co-founded by industry leaders, galvanized by successful entrepreneur and investor, Saurabh Srivastava (who was, incidentally, also involved in co-founding NASSCOM). Major investment institutions, private equity funds and companies are actively engaged with the platform.

Globally, angel groups are usually informal, single-city operations, meeting once in a couple of months and making four or five investments in a year. On the other hand, IAN's operations are spread across six cities in India (Delhi, Bengaluru, Mumbai, Pune, Hyderabad and Kolkata) and one chapter in London (launched by Prime Minister David Cameron). This makes IAN the world's first angel group to be multi-location as well as global. This allows IAN to find the most innovative ventures in India and overseas and help them

8. Information on IAN from www.indianangelnetwork.com and personal inputs from Mr Saurabh Srivastava.

scale and grow across the country and overseas. It now has regular pitch sessions in London as well (in addition to the six Indian cities), bringing innovative British ventures to not only raise funds at IAN but also leverage the large global market access that IAN provides to build global footprint companies. Several innovative British companies have now joined the IAN portfolio and are rapidly scaling up across borders including India, Singapore, the US, etc.

Annually, close to 5,000 entrepreneurs reach out to IAN and undergo a quick but deep multilayered diligence process, which shortens the list to three hundred ventures. These ventures then present to IAN investors through weekly pitch sessions. The strength of the Network and its members' faith in the diligence and curation results in around forty investments annually, with the typical US$ 500,000 investment committed to a company within twenty-four hours! IAN takes the engagement beyond just helping companies raise funds; the entrepreneurs are not only mentored by investors, but IAN supports them in accessing global markets and clients, developing their teams, building their visibility and brand, raising the next round of investments, etc. This post-investment support and engagement helps the companies to grow their businesses faster and build globally competitive companies.

With a portfolio of over eighty companies invested across seven countries and multiple sectors (seventeen sectors), IAN has met with early successes. A number of its marquee investee companies have given investors outstanding returns with a potential to become unicorns. IAN's run rate of investments is now Rs 100 crore (about US$ 15 million) per year, and its investee companies typically raise five times that amount in the next round.

One of the interesting incubators is Startup Village (SV) in Kochi, Kerala.[9] It is a unique public-private partnership (PPP) project that has created a model incubator through the collaborative efforts of the Government of India, Technopark (an IT park set up by the state of Kerala) and MobME Wireless, a private tech company. It is a not-for-profit society, focused on the themes of mobile and Internet, and supports young founders of start-up companies so as to enable them to realize their potential of becoming entrepreneurs through the journey from 'idea to IPO'. In less than three years from its inception, it stimulated the start-up scenario in the country, igniting the dormant entrepreneurial instincts in young minds, and turning them into job creators rather than job seekers.

SV has an interesting genesis. In 2006, Sanjay Vijayakumar and his friends, college students at that time, approached Technopark, the largest IT park in India, to set up a technology start-up. There was no technology incubator at Technopark, but using this student start-up as a case study, Technopark approached the National Science and Technology Entrepreneurship Development Board of the Government of India to set up a Technopark technology business incubator (TBI). Since Sanjay's start-up (MobME) was then the first and only campus start-up, the officials included Sanjay on the board of governors to help streamline the functioning of the incubator.

Along with the growth of MobME from a campus start-up to being recognized as one of India's 'Top 10 Emerging IT Companies', Technopark TBI also grew rapidly and by

9. Details on Startup Village draw on conversations with and inputs from Mr Sanjay Vijayakumar.

2012 it had more than 150 start-ups. The defining moment
came when it won the President's Award for the best TBI in
India. However, the growth also brought challenges in speed
of execution and flexibility, given the constraints that were
ingrained in the government structure.

To sustain the growth and momentum of the incubator,
MobME and Technopark approached DST to set up a unique
public-private partnership to bring together the execution
skills and flexibility of the private sector and the strength of
the public sector. The idea was to create an incubator that was
credible and had the capacity to take large risks for public
good.

In six years, Sanjaya and his team had built MobME into
a company with a market cap of Rs 100 crore. Based on their
own experience, they realized that the start-up ecosystem in
India was very weak when compared to Silicon Valley, though
the ability of Indian youth was as good as that of their peers
in Silicon Valley. This led to a situation in which many Indian
entrepreneurs moved to countries where congenial and efficient
ecosystems were available. Sanjay decided to take up the
entrepreneurial challenge to build a start-up ecosystem where
India could also produce billion-dollar start-ups from college
campuses, similar to Apple, Microsoft, Google or Facebook,
which were all campus start-ups.

This defined the objective of Startup Village: to create an
ecosystem for a thousand product start-ups in India and to
catalyse a billion-dollar start-up from a college campus. Its
mission is to establish India amongst the top five start-up
ecosystems in the world and thus play a significant role in the
creation of world-class policies, infrastructure, incubators,
human capital development programmes and angel funds to
create knowledge, employment and wealth in India.

The role of Technopark and MobME as joint host institutes is also clearly defined. Technopark is responsible for administrative support and provision of infrastructure while MobME Wireless is responsible for the vision, execution and management of the incubator. MobME is also responsible for raising the funds required to build a world-class start-up ecosystem in India so that its young talent stays in the country, instead of moving to Singapore or Silicon Valley.

Startup Village seeks to create a conducive ecosystem for start-up companies. Obviously, policies that will put India amongst the best places in the world to establish a start-up (and improve its 2015 rank—155th of 189 countries—in the ease of starting a business in the World Bank index[10]) are necessary. Startup Village provides physical incubation services that include furnished co-working spaces, power, water, security, high-speed Internet, access to angel investors and technology partners. Common infrastructure such as high-end computers and devices, which are expensive, are also given to the start-ups at a nominal rent. In addition, it provides access to top mentors.

Startup Village companies, unlike ventures located elsewhere, have favourable tax laws (for example, companies having a turnover of up to Rs 5 million are exempted from service tax). It operates an in-house angel fund that start-ups can tap into.

Technology Innovation Zones set up by partners are a unique facility at Startup Village. The Power Innovation Zone (set up by the Kerala State Electricity Board [KSEB]) is a good example; here, start-ups can pitch their innovative

---

10. 'Ease of doing business: India improves ranking, Singapore tops the list, says World Bank', *The Economic Times*, 28 October 2015.

ideas directly to KSEB. Companies such as Blackberry, Federal Bank, Microsoft, Oracle and IBM are other partners in this programme.

Startup Village focuses a lot of its energies on developing the human capital pipeline. This is done through hundreds of workshops held each year on various themes such as robotics, aerospace, Android and iOS development. These workshops are held both at Startup Village as well as in colleges.

Startup Village sees itself as being more than just a business incubator; it wants to change the way Indians look at entrepreneurship. Their education campaigns and initiatives aim to provide students the skills they need to design world-class technology products and start their own companies. Startup Village conducts boot camps in engineering colleges; has established a leadership academy to impart leadership training to youth; collaborated with FAB Labs from MIT, Boston, to build hardware products; started a Raspberry PI programme for eighth-standard schoolchildren; introduced Startup Box for college students; and set up a 'landing pad' in Silicon Valley. In January 2013, Startup Village (SV) and the Kerala government launched an initiative called SVSquare, with the second 'SV' being Silicon Valley (interestingly, the initials of the initiator too are SV). Every year, promising young entrepreneurs from India are selected and sent on an all-expenses paid trip to the US. The aim is to expose the youngsters to the legendary start-up environment in Silicon Valley.

A community gathering is held on the third Saturday of every month in the Startup Village campus in Kochi and the latest products developed there are showcased to prominent entrepreneurs, investors and mentors who visit from all over the country. This enables students and founders to hobnob with some of the biggest names in India's business community, learn from the experience of well-known entrepreneurs and connect with investors. By 2016, over 6,000 entrepreneurs had applied to Startup Village and 800 start-ups have been set up already, of which as many as 276 are student start-ups. The Student Entrepreneurship Policy of the Kerala government provides for 20 per cent attendance and 4 per cent grace marks every semester to student entrepreneurs. This path-breaking initiative is a first in India. In another defining policy, Kerala now spends 1 per cent of its state budget on youth entrepreneurship.

The success of Startup Village in Kochi has kindled interest in this ecosystem model in other states, which would like to emulate this model. The newly formed Andhra Pradesh (following the bifurcation of the earlier eponymous state) is keen to create a similar facility in Visakhapatnam. Headed by the dynamic and IT-savvy chief minister, Chandrababu Naidu, the state has already signed a Memorandum of Understanding with MobME to set up a Startup Village in Andhra Pradesh. The chief minister has also released India's first start-up policy, taking the cue from the report on skilled youth that Sanjay had prepared for the State Planning Board. Many states are likely to adopt this model to create a culture of entrepreneurship amongst students.

At a broader level, NASSCOM—the association of IT software and services companies in India—provides an example of the role of a sector-wide player in promoting innovation through entrepreneurship. For many years, it has been encouraging start-up companies through a variety of initiatives: workshops, round-tables, regular small-group meetings with volunteer-mentors, market intelligence reports, awards and showcasing opportunities, meetings with potential investors, etc. These are now being drawn together in a '10,000 Start-ups' programme, aimed at creating that number of new companies in the next few years.

This start-up programme is being conducted with support from industry, academia, government, incubators and angel funding groups.[11] A hundred-seater 'Start-up Warehouse' was established in Bengaluru in August 2013. This provides world-class co-working space and incubation facilities. The NASSCOM programme had received 9,000 applications by December 2014. Of these, 2,500 were screened, 800 were shortlisted and over 200 were either funded, mentored or incubated. To encourage start-ups and spread the message, over 350 events were conducted in fifteen different cities.

NASSCOM estimates that in 2014 there were 3,100 start-ups—triple the number that existed in 2006. The distribution of these is as expected, with a majority in Bengaluru (28 per cent) and Delhi/NCR (24 per cent). Mumbai has a much lower share, at 15 per cent, followed by Hyderabad (8 per cent), Pune (6 per cent) and Chennai (6 per cent). To the extent that most start-ups must have an innovative component (product, process, business models), the geographical distribution could be considered a good surrogate for innovativeness of the particular city. However, there are many cases where start-ups begin in one city but soon migrate to another (Bengaluru, in particular), due to the availability of a better ecosystem or specifically for talent.

India is now recognized as a global hub for innovation and start-ups. In 2015, its 4,200 to 4,400 start-ups placed it at the third rank, just below the UK, though an order of magnitude

---

11. Details and data are from NASSCOM.

fewer than the US number. Venture capital and private equity have flowed generously to Indian start-ups, totalling over US$ 2.2 billion in 2014 and an estimated increase to US$ 5 billion in 2015. Not surprisingly, most founders of start-ups (73 per cent) are below thirty-five years of age, with 15 per cent being even below twenty-five[12].

Interestingly, start-ups are receiving funding and mentorship from successful recent entrepreneurs. Some successful entrepreneurs are also providing start-ups with free workspace in their premises.[13] For example, Paytm—a very successful, though as yet young, entrepreneurial venture in the field of online payments—provides space to a number of entrepreneurs, and as many as twenty-four start-ups have been incubated on the Paytm premises. Vijay Shekhar Sharma, founder of Paytm, sees it as 'my way of giving back to the ecosystem'. Such fraternity amongst entrepreneurs is bound to create a more favourable ecosystem, giving a leg up to entrepreneurship and innovation.

Initiatives such as the 10,000 Start-ups programme are helping to give a tremendous boost to entrepreneurs and to innovation. India seems to be poised to make a quantum jump in this area. It certainly has many of the characteristics that provide the right ambience for innovation. One would, therefore, expect it to be amongst the premier countries in any ranking of innovative nations. The following chapter looks at where India stands in this regard.

---

12. NASSCOM report as quoted in 'Indian start-ups to see funding worth $5 billion by year end: Nasscom', *Mint*, 13 October 2015.

13 Many media articles have covered such stories. This example is from 'Meet the brotherhood of entrepreneurs', *The Economic Times* (Delhi), 8 September 2015.

4

WHERE DOES INDIA STAND?

India's position in global innovation rankings; its poor performance in quality of school education and in university rankings. Importance of university–industry interaction. Talent as a driver, immigration as an enabler (examples of Israel, US, Singapore); supportive government policies.

Today, countries around the world recognize that we are in the midst of another major transition in terms of the determinants of power and prosperity. In prehistoric times of hunter-gatherer societies, power and economic well-being were defined by the skill, numbers and weapons of the hunters. The move to settled agriculture made land and water the key resources. While these remain important, financial resources assumed increasing importance in the industrial age, with capital becoming the measure of power for individuals, communities and countries. Through much of the twentieth century, capital remained dominant, but technology rapidly emerged as another vital component of power. Countries with high levels of technological capability could not only produce better products, but did so more efficiently. Technological know-how, guarded by intellectual property rights, itself became a tradeable commodity of high value.

Now, as the knowledge economy has begun to mature, we are witnessing the emergence of innovation as the new driver of economic growth. Information and knowledge, bundled together as technology, is taken forward by innovation. A large industrial base and manufacturing infrastructure are no longer the signs of a powerful country. In fact, given the concerns and high public awareness about the dangers of pollution due to industrial activity, many countries, particularly developed ones, prefer to outsource industrial activity to an appropriate offshore destination in another country. As a result, the currency of power is now innovation—the capability to conceive and design a new product, or come up with a new idea. If the 'currency of power' is itself an outmoded idea that needs to be thought of more innovatively, an option would be 'power of currency'. In a globalized and increasingly interdependent world, the strength of a country's currency is a fair indicator of the economic power of that country as assessed or perceived by others. While short-term spikes and valleys may occur, the long-term trajectory of a country's currency is dependent on its perceived capabilities with regard to technology and innovation.

With this perspective, it is but natural that country after country is now on the innovation bandwagon. One indication of this—as also a catalyst for it—is the number of books, reports and rank lists on innovation. The various approaches, viewpoints and criteria provide a fair diversity, though in most cases, the group of countries considered most innovative is broadly similar.

One report with a high standing is the yearly Global Innovation Index (GII) published by World Intellectual Property Organization (WIPO). *The Global Innovation*

Index 2013: The Local Dynamics of Innovation is the result of a collaboration between Cornell University, INSEAD and WIPO as co-publishers, and their knowledge partners. The index uses a framework that looks at innovation inputs and outputs, and defines an innovation input sub-index and an innovation output sub-index. The former is based on five 'pillars' and the latter on two. Each pillar is further subdivided into three elements. The two sub-indices are combined into an innovation efficiency ratio, which is then used to derive the Global Innovation Index, on which the overall rank of each country is based.

The 2013 rankings put India, with a score of 36.2 (on a 0-100 scale), at number 66 amongst the 142 countries that were assessed. The top rank went to Switzerland (with a score of 66.6), followed by Sweden, the UK, the Netherlands and the US. Hong Kong and Singapore also figured in the top ten (at ranks 7 and 8 respectively). Israel, often considered a star in innovation—especially in the IT sector—was at rank 14.

The GII also ranks countries by their standing within their cohort. In the grouping of thirty-six lower-middle-income countries, India was ranked third (after Moldova and Armenia). The top country within this category (Moldova) was at rank 45 in the overall list. The forty-four countries preceding it were all from the categories of high-income and upper-income economies. Similarly, the top-ranker (Malaysia) in the upper-middle-income category was at overall rank 32, with all the countries above it being from higher income categories. Does this indicate a correlation between income and innovation? More interestingly, does higher income lead to greater innovation, or vice versa? In another chapter, it is argued that deprivation is a major driver of innovation, especially at the grass-roots level. Yet, according to the GII, the top ten countries (in fact, the top thirty-one) in overall ranking were all high-income economies. This does point to a definite link between economic prosperity and innovation.

The sub-index on innovation output presents a different picture, though. Moldova, ranked 28 on this sub-index, did better than nine of the top ten upper-middle-income countries. In terms of the innovation efficiency ratio, Mali—a low-income economy—topped the list. Thus, in this case, all the 121 countries in higher income groups were ranked below it. Therefore, a more granular analysis belies the innovation–income link. This is certainly encouraging, especially for the lower-income economies, as it opens the possibility of their competing on the innovation front even before their economies grow to more prosperous levels. For India, with endowments and characteristics that can take it to the top group in innovation, this is a significant insight. It could well be the country that charts the way in terms of using innovation as a driver of economic growth.

While India undoubtedly has great potential, its low rank (66) in the GII placed it poorly even amongst its peers. For example, if we look at BRICS, Brazil and Russia were ahead, even if only marginally, at ranks 64 and 62 respectively; China had a substantially higher ranking at 35, while South Africa was at 58. Thus, India was at the bottom of the list in this grouping. Apart from China, India did not come out too well in comparison with other Asian countries either. Japan, once famed for being a leader in cutting-edge electronics and the originator of many innovative products, and despite its long economic slowdown, was yet ranked 22 in the GII. Quite expectedly, the new Asian star, South Korea, did well and was at 18. Malaysia, a country many would consider new to the game, even an upstart, was way ahead of India and was ranked 32. Surprisingly, countries hardly known for innovation were also ahead of India—for example, Saudi Arabia (42) and the Gulf countries, United Arab Emirates (38), Qatar (43) and Kuwait (50). As noted earlier, Hong Kong and Singapore were ranked far above India, figuring in the top ten.

All the South Asian countries fared poorly in the GII. Pakistan was pretty much at the bottom of the heap at 137, near-neighbours Nepal and Bangladesh were nearly so too, at 127 and 129 respectively, while Sri Lanka—the star performer in various social indices—was at a lowly 98. One wonders why this poor ranking of all South Asian countries. An understanding of this can come only through a detailed analysis of the seven pillars and their sub-elements, which provide the inputs that determine ranking.

In India's case, its ranking in innovation output was far better than that in innovation inputs (expectedly, it topped the world ranking in exports of computer and information services, and was ranked 11 in export of creative goods). On the input side, it scored poorly with regard to institutions and human capital. One result of being ranked low in input as compared to output (rank 87 versus 42) was that India was ranked as high as 11 in innovation efficiency.

Looking at the future, what is particularly disquieting is India's low rank (90) under the head 'Expenditure on Education'. Though based on current expenditure on education as a percentage of gross national income (which should make it size-agnostic), the ranking seems, for some reason, to favour small countries (eight of the top ten countries were developing countries, and all, barring Ghana, were small). Yet, Singapore and Hong Kong—both in the top ten of overall GII ranking—were even below India (at 92 and 98). Low expenditure on education would, in theory, impact the human capital base and hence a country's future ability to carry out research and development and to innovate. One could, of course, make a conscious strategic choice to invest less in education and make up for that by actively promoting immigration of highly skilled people into the country. However, it is unlikely, for a variety of political, sociological and strategic reasons, that any country is actually doing this.

Many feel that India has, over the last few years, been recognized for its capability in innovation. This may be so, but the figures tell a different story. India's already low overall rank (66) in the GII in 2013 actually fell further: to 76 in the 2014 assessment, and further to 81 in 2015. This marked a continuous downward slide from 2011: from 62 to 64 to 66 to 76 and in 2015 a big drop to 81.[1] The good news is that the 2016 GII shows a very big bounce-back, propelling India to rank 66.[2]

Another education-related indicator is school-life expectancy, a measure of how many years of education a child of school-entering age would receive during his or her lifetime if the school enrolment rates stay the same as of today. In 2013, India ranked a lowly 109 on this. Apart from extent of schooling, the quality of schooling too is an issue. For this, the Programme for International Student Assessment (PISA) is an

---

1. Figures from a discussion with Prof. Soumitra Dutta (an author of the report) in Delhi, June 2015.

2. 'Global Innovation Index: India moves up to 66th rank this year', *The Economic Times*, 17 August 2016.

internationally accepted benchmark for assessing the capability of school students in reading, mathematics and science.[3]

PISA is a worldwide study by the Organisation for Economic Co-operation and Development (OECD), in member and non-member nations, of fifteen-year-old school pupils' scholastic performance in mathematics, science and reading. It was first conducted in 2000 and then repeated every three years. It is done with a view to improving education policies and outcomes.

PISA is a large-scale survey, with as many as 470,000 fifteen-year-old students representing sixty-five nations and territories participating in 2009. An additional 50,000 students representing nine nations were tested in 2010.

PISA 2012 assessed the competencies of around 510,000 students between the ages of 15 years and 3 months and 16 years and 2 months in sixty-five countries and economies. In forty-four of those countries and economies, about 85,000 students also took part in an optional assessment of creative problem solving. This paper-based two hours' test was a mix of open-ended and multiple-choice questions that were organized in groups based on a passage setting out a real-life situation. Students and school principals also answered questionnaires to provide information about the students' backgrounds, schools and learning experiences and about the broader school system and learning environment.

PISA 2012 was presented on 3 December 2013, with results for 510,000 participating students in all thirty-four OECD member countries and thirty-one partner countries. This testing

---

3. PISA 2012 Results in Focus, Programme for International Student Assessment (PISA), Organisation for Economic Co-operation and Development (OECD).

cycle had a particular focus on mathematics. Shanghai had the highest score in all three subjects. It was followed by Singapore, Hong Kong, Chinese Taipei and Korea in mathematics; Hong Kong, Singapore, Japan and South Korea in reading; and Hong Kong, Singapore, Japan and Finland in science.

India's rank in 2010 was a dismal 69 out of the 70 countries ranked, while China was ranked 1. Hong Kong followed at 2, while Singapore, Korea and Japan were ranked 4, 5 and 6 respectively. While this ranking dates to 2010 for India (since then India has not participated) and 2009 for others, there is no reason to assume any radical change since then.

Similar results, indicative of the poor quality of education, come from another large Indian survey, the Annual Status of Education Report (ASER; the acronym means 'impact' in Hindi). ASER is the largest non-governmental household survey undertaken in rural India and is facilitated by the Pratham Education Foundation.

True to its meaning, the survey measures the enrolment status of children between the ages of 3-16 years and tests their basic reading and arithmetic abilities through a detailed process that uses a common set of testing tools and a comprehensive sampling framework.

ASER has been conducted every year since 2005 in all rural districts of India. In 2013, ASER reached about 600,000 children in more than 16,000 villages in 570 rural districts of India. Around 30,000 volunteers and 500 partner organizations, including educational institutions, participated in the survey.

The findings from the ASER surveys have indeed been disappointing. The 2014 report[4] indicated that learning levels

---

4. Annual Status of Education Report (Rural) 2014—Provisional Results.

across the country, whether in public or private schools, had not improved. The percentage of Standard III children able to solve simple two-digit subtraction problems fell from 26.1 per cent in 2013 to 25.3 per cent in 2014. The percentage of children in Standard II who could not recognize numbers up to nine increased from 11.3 per cent in 2009 to 19.5 per cent in 2014. Only 48.1 per cent of Standard V children were able to read a Standard II-level text in 2014—a small increase from 47 per cent in the previous year. Meanwhile, enrolment in rural India (of six- to fourteen-year-old children) remained stagnant at 96.8 per cent, and attendance in both primary and upper primary classes showed a steady downward trend. In 2009, attendance was 74.3 per cent in primary schools and 77 per cent in upper primary schools as compared to 71.4 per cent and 71.1 per cent respectively in 2014.

The silver lining has been the improvement in physical infrastructure: more schools with toilets—and importantly, more separate toilets for girls (55.7 per cent in 2014, compared to 32.9 per cent in 2010)—better pupil–teacher ratio, improvement in drinking water facilities and a small increase in the proportion of schools with libraries.

A striking feature over the last few years has been the growth in private schools. While these accounted for only 16 per cent of the enrolment in 2005, the figure had almost doubled to nearly 30 per cent in 2014. Present trends indicate that this number will increase to 50 per cent by the end of this decade. This points to the willingness of parents to pay for education, but it also reflects the perception of poor quality of education in government schools. These have seen a decline in the absolute number of students, which dropped by 11.7 million, from 133.7 million in 2007–08 to 121 million in

2013–14 (including both primary and upper primary). This period saw enrolment in private schools rise by 27 million, to 78 million.

For many years now, successive governments in India have set a target of allocating 6 per cent of the country's gross domestic product (GDP) for education. In 2004, the then-incoming government reiterated this target. However, even in 2012, the figure was only 4.2 per cent. Over the years, and especially from 2005, there has been a big expansion at all levels of education, from primary to tertiary. The Right to Education law (assuring all children schooling up to Standard VIII) and many new institutions of higher education have meant a big quantitative growth in education. There are, however, concerns about whether this would mean a further dilution of quality. While PISA, as also the ASER study, points to the poor learning in schools, there is concern about higher education too.

The Indian higher education system is amongst the largest in the world in terms of the number of colleges and universities. From 350 universities and 16,982 colleges in 2005–06, the numbers increased to 713 universities, 36,739 colleges and 11,343 diploma-level institutions in 2013–14. The gross enrolment ratio (GER) in higher education nearly doubled from 11.6 per cent in 2005–06 to 21.1 per cent in 2012–13 (provisional), with 29.6 million students enrolled in 2012–13 as compared to 14.3 million in 2005–06.[5]

Despite (or possibly because of) this rapid expansion, quality has been a casualty. While there is no common benchmark (such as PISA or ASER for the school level), the quality of

---

5. Economic Survey 2014–15 (Government of India), February 2015, Volume II.

the average graduate is generally rated as being poor. Despite the fact that some Indian institutions do produce world-class graduates and are widely respected (for example, the Indian Institutes of Technology), not a single Indian university has figured in the top two hundred in global rankings of well-regarded assessments.

In the widely recognized Times Higher Education World University Rankings of 2014–15,[6] there was again no Indian university in the top two hundred. The first institutions from India figured in the 276-300 bracket, which had the well-known Indian Institute of Science (IISc) and also (the less widely recognized) Panjab University. The latter was in the 226-250 group in the previous year, and so had slipped in the rankings and does not figure in the top 500 in the subsequent two years. IISc, on the other hand, improved its ranking and is in the 201-250 in 2016–17.

Two other Indian universities figured in the top four hundred in the 2014–15 ranking: the Indian Institute of Technology (IIT) Bombay and IIT Roorkee, both in the 351-400 group. IIT Delhi, IIT Kharagpur and IIT Kanpur, which were all in the 351-400 band in the previous year, were no longer in the top four hundred. The 2016–17 ranking has IIT Bombay yet in the 351-400 band. IIT Delhi, IIT Kanpur and IIT Madras are all in the 401-500 group.[7]

The rankings are compiled on the basis of thirteen categories across five dimensions—teaching, research, citation impact,

---

6. Times Higher Education World University Rankings of 2014–15.

7. Data for the 2016–17, Times Higher Education (THE), The World University rankings 2016 are from *The Times of India*, 22 September 2016.

international outlook and industry income. Obviously, given their ranking, Indian universities did not do well in any of these dimensions, but they fared particularly poorly in 'international outlook'. While one might question the adequacy and relevance of the criteria, the rankings are doubtless a good index of the comparative international standing of Indian institutions of higher education.

Indicative of the malaise in education is India's ranking in the global Human Capital Index,[8] which measures countries on development and deployment of human capital. This ranking by World Economic Forum looks at forty-six indicators, focusing on education, skills and employment. India was ranked 100 in the 124-nation list in 2015, placing it not only lower than its BRICS peers, but also below neighbours like Sri Lanka and Bangladesh.

Just as education is the foundation for human and economic development, investment in research and development is a critical determinant of technological prowess and a catalyst for innovation. India has, for long, aimed at a target of 2 per cent of GDP to be invested in research and development but even in 2016, this goal had not been achieved. As a result, India compares poorly against other aspirant countries. The Global Innovation Index provides one view of this. Comparative data across countries for gross expenditure on research and development as a percentage of GDP (for 2009; although dated, there has probably been little relative change) put India at rank 43—below the other four BRICS countries. At the top, not surprisingly, was Israel, while South Korea, known for its focus on high-tech, was ranked 3. China was ranked 21

8. *The Human Capital Report 2015*, World Economic Forum.

in the list and has probably moved up subsequently, given its growing focus and expenditure on research and development.

Another index of innovation is the number of patents filed, indicative of the extent and depth of the invention-innovation activity. According to a news report,[9] 39,400 patent applications were filed in India in 2010, along with 7,589 design applications. In the same year, 490,226 patent applications were filed in the US; 391,177 in China; 344,598 in Japan; and 170,101 in Korea. The numbers for China and Korea are particularly noteworthy. These two countries are probably at the forefront of those seeking to use technology and innovation as the future driving forces of their economies.

It is true that many inventors in India do not file patents, either because of limited awareness of the importance of doing so, or the comparatively high cost. Despite this, the numbers quoted here are indicative of the huge gap between India and some of the leading countries in this area.

In some countries—the US in particular—companies often file multiple patents for a single product so as to ensure protection for every element of intellectual property (IP) in it and to create entry barriers for competing products.

As an example, consider the Gillette Mach3, a line of safety razors produced by Gillette and introduced in 1998. It was the world's first triple-blade razor, and involved more than US$ 750 million in research and development costs. Gillette spent US$ 300 million in the first year to promote it.

Razors are among the most heavily patented consumer products, with more than a thousand patents covering everything from lubricating strips to cartridge-loading systems.

---

9. *The Economic Times*, New Delhi, 28 March 2014.

Gillette has more than fifty patents covering its Mach3 franchise. Thus, a single item of everyday use which, on the face of it, does not appear to be a complex device invented by some high-tech laboratory, is the product of a huge US\$ 750-million research-and-development effort and has as many as fifty patents protecting it![10]

Innovation, as noted earlier, takes place in a number of diverse fields. It is by no means limited to the arena of science and technology. Yet, given the preponderant and growing role of science and technology, it is obviously a major force that propels innovation and new products. A country's capacity in science and technology is therefore of considerable importance. One indicator of the robustness and productivity of a country's science and technology infrastructure (human and physical) is the number of publications by its scientists in respected or refereed journals. While the number of publications from India has been increasing, it is at a rate far lower than, for example, China. As a result, India's share in such publications has, in fact, been decreasing.

We have noted earlier how cities like Las Vegas and Dubai have been innovative in repositioning and rebranding themselves. Similarly, countries like Singapore and Israel have innovatively reinvented themselves. Singapore has transformed itself from a somewhat sleepy though well-located port to a major trans-shipment hub (for goods as well as air travellers), an efficient city, a financial hub and trading centre. It has ambitions now to be an important centre of culture and a knowledge city. It seeks to attract the best talent from around the world and is facilitating the setting up of branches of major

---

10. Details on the Gillette Mach3 are drawn from 'The war of the razors', *The Boston Globe*, 31 August 2003.

global universities, so as to become a vibrant centre of research. It has, through the last few decades, constantly innovated and come up with new identities that it continues to evolve.

The transformation of Israel has been equally, if not more, dramatic. Created, many would say, to assuage the West's guilty conscience and amidst Zionist terrorism, the country had few natural resources. Arid lands, a scarcity of fresh water and a large immigrant population were serious problems. As if these were not enough, the young country also faced a hostile security environment, threatened by neighbours who resented the creation of a new state by the division of their land. Going through the kibbutz phase of collectives—not only for farming, but also for raising the young—it soon became an example of resilient agriculture. Innovative techniques like drip irrigation helped minimize the use of that valuable and scarce resource, water, and made the desert bloom.

Today, Israel is seen as one of the epicentres of high technology. Many young start-ups, conceived, born and incubated in Israel, have launched huge and very successful IPOs on Wall Street. Unlike Silicon Valley, where the government plays a small or non-existent role, the Israel story has the government as an active and key player. Crucial start-up finance and, equally important, a conducive ecosystem, can both be attributed to the government. As in the US, a key element of success has been the collaboration between academia (and/or research institutions) and the entrepreneur.

A major instrument of the Israeli government in promoting tech-entrepreneurship and innovation is the Office of the Chief Scientist (OCS).[11] It is specifically charged with the

---

11. Overview of OCS and MATIMOP are drawn from the official website of MATIMOP, Israeli Industry Centre for R&D

responsibility of fostering the development of industrial research and development, and its mission has been defined through the 'Law for the Encouragement of Industrial Research and Development—1984'. A research and development fund and a wide range of international programmes, agreements and collaborations facilitate its work. OCS aims to assist the advancement of Israel's knowledge-based science and technology industries in order to encourage innovation and entrepreneurship while stimulating economic growth. It does this through a number of programmes which provide financial and developmental resources for industrial R&D within Israel to entrepreneurs and companies of all sizes at various stages of growth. This is a recognition of the fact that the development of innovative new commercial technologies, products and services is a high-risk and expensive proposition. Since this often exceeds the capacity and capabilities of individual firms, OCS steps in to provide support. Where there is need for international cooperation, OCS facilitates this too.

MATIMOP serves as the executive agency of OCS and is the official agency for industrial R&D cooperation in Israel. It formulates policies to support and promote Israel's industrial infrastructure, by nurturing industrial innovation and entrepreneurship. It also develops and implements international cooperative industrial R&D programmes between Israeli and foreign enterprises.

Since much of its global markets (especially the US) are far from Israel, it uses international agreements to bridge the distance. It has charged MATIMOP to engage in mutually beneficial international collaborations so as to aggressively expand opportunities for Israel's industry.

The remake of Israel surely has lessons for countries aspiring to tap into technology and innovation as the drivers for economic growth. Some factors may seem unique to Israel: for example, the inflow of a large number of highly qualified professionals, particularly from Russia and East and Central Europe. Yet, even this is capable of being emulated. In fact, some countries actively encourage the immigration of technology professionals, with easy entry, quick long-term residency permits, etc. Singapore and the US are amongst the countries that do this. The payoff is visible— though actual estimates vary, it is widely acknowledged that a large proportion of the start-ups in Silicon Valley (including many high profile ones) were founded by immigrants. Amongst immigrants whose names are widely recognized are Sergey Brin, Andrew Grove, Vinod Khosla and Sabeer Bhatia.

In 1999, the dean of UC Berkeley School of Information, AnnaLee Saxenian, discovered that Indian-born entrepreneurs had founded 7 per cent of all Silicon Valley start-ups between 1980 and 1998.[12] In 2007, nearly eight years after Saxenian published her findings, Professor Vivek Wadhwa partnered with her and Professor F. Daniel Siciliano of Stanford Law School to update and expand the research. The results were astonishing. Twenty-five percent of the nation's start-ups and 52 per cent of those in Silicon Valley were founded by immigrants. Indian immigrants were the leading company-founding group. They founded 13.4 per cent of Silicon Valley's start-ups and 6.5 per cent of those nationwide. This was particularly surprising,

---

12. AnnaLee Saxenian, 'Silicon Valley's New Immigrant Entrepreneurs', *Public Policy Institute of California*, 1999.

because Indian immigrants comprised less than 1 per cent of the US population at the time.

There is clearly a lesson here for countries like India, where immigration laws are tight and there is no special encouragement provided even to high-tech foreign professionals who want to make India their home.

Another well-known driver of innovation is the interaction between the entrepreneur and researchers or academics. At a broader level, one can look at this in terms of the depth of interaction between industry on the one hand, and universities or research institutions on the other. The closeness and intensity of this interaction is often the crucial accelerator to translating a new idea into an innovative product. As seen in examples from around the world, physical proximity is important—even in today's distance-less world—and it is therefore not surprising that the major concentrations of innovation are found around major universities and research institutions. Often, incubators or accelerators located within the university, or nearby, have been an important factor in encouraging and fostering innovation. This once again demonstrates the advantages of physical proximity.

Innovation often takes place at the boundary between disciplines, or by cross-pollinating ideas or analogies from one discipline to another. This requires that persons from various disciplines interact with each other, and what better place for this to happen than at a university? Even better: not just one, but a clutch of universities—a 'university town'. It is for this reason that so many of the globally renowned start-up centres are in such multi-university education and research centres. Clearly, this provides another possible point of policy intervention: the need to create clusters of educational and research institutions, rather than dispersing them in a sub-critical way.

The top rankings in the Global Innovation Index (referred to earlier) seem to be skewed towards smaller and comparatively prosperous countries: the only large country in the top ten was the US (and the UK, if one wants to consider it as large), and the top thirty ranks were all taken by high-income countries. Arguably, innovation and prosperity are not independent variables—i.e. one does influence the other—but what about size? It seems plausible that a smaller country results in a concentration or higher density of institutions, facilitating the trans-disciplinary interaction that triggers innovation. Even in a large country like the US, it is in areas of institutional high density (for example, in the Northeast, or around San Francisco and Silicon Valley) that one sees a much higher intensity of innovation. The high density, one could hypothesize, automatically leads to inter-institutional interaction—formal or informal—and, inevitably, collaboration. It also increases the probability of closer interaction between these institutions and industry or entrepreneurs. Together, these factors result in more innovation.

The poor record (academic and research and development) of institution-industry interaction in India affects its ranking in the Global Innovation Index, and is also a cause for concern with regard to future innovation. Yet, India is amongst the countries well placed to be global hubs for innovation. India's advantage derives from its diversity and democracy, as also the context of adversity: each a driver of innovation. The conventional perspective of a risk-averse middle class needs drastic revision in the light of the experience of the last few decades, especially in the sphere of technology. Little wonder that India is today a major and thriving centre for start-up ventures (some of the initiatives were mentioned in the previous chapter), many of which are based on innovations in product, process or business model.

This is particularly important for India, because its requirement for massive job creation can best be fulfilled by new enterprises, as experience elsewhere has shown. Its competitiveness in the global marketplace too will depend upon innovation, as it is unlikely that India will be able to compete with others in low-cost, low-value goods (given its poor cost-productivity and inefficient infrastructure). Even in services, the cost-quality-delivery mantra of its IT software and services industry will have to be supplemented with a new differentiator: innovation.

Creating the right policy framework and ecosystem for competitiveness and innovation is therefore critical to the country's future. In this context, a new initiative is noteworthy. As part of the government's effort to stimulate innovation and research so as to promote domestic high-value manufacturing, school students (at the class nine level) could soon be learning about intellectual property rights. The concerned officials from the government are in talks with the National Council for Educational Research and Training (NCERT) and the Central Board of Secondary Education (CBSE) to introduce intellectual property rights in the curriculum from class nine. The government is also looking at introducing intellectual property rights as a subject in engineering colleges. At present, about 40,000 patents a year are filed in India (42,774 in 2014-15), as compared with about 500,000 in the US.[13]

---

13. Based on a report in *The Economic Times* (Delhi), 21 May 2015.

While the role of the central government is vital, a lot of the action will necessarily have to take place at a more decentralized—state and city—level. In what follows, we discuss these aspects and evaluate the relative advantages of various states and cities of the country.

5

CREATING INNOVATION HUBS

The relative competitiveness of Indian states. Socio-cultural and other factors that determine innovativeness and entrepreneurship in a state. What makes cities innovative?

Competition among states in India is comparatively new and is yet rather limited. Primarily focused on attracting private and foreign investment, such competition really began in a limited way from around the end of the 1990s. At first, only a few states were involved, with Andhra Pradesh and Karnataka being the most active. The focus was on the IT software industry, partly because this was seen as a highly visible, glamorous and sunrise sector, with a high growth potential.

There were, of course, also good, hard-headed reasons to focus on IT. First, it was seen as a high employment generator, especially for urban youth. Politically, this was important because it not only created jobs, but provided them to educated and mainly urban youth: a vocal and volatile group whose disaffection could potentially spell trouble for any government. Incidentally, it also created many jobs for semi-skilled or unskilled people in support services like security, housekeeping and transportation. In addition, the industry needed large amounts of high-quality office space, triggering a construction boom that benefitted other industries even as it generated more employment. Finally, the high wages it paid, mainly to young people, meant a lot of disposable income. This led to a boom in demand for housing (since many employees were migrants), domestic help, cars and motorcycles, restaurants, entertainment, etc. The economic and employment spins-offs were thus quite substantial.

Bengaluru, with its conducive ecosystem (more on this later), salubrious climate and cosmopolitan culture, was a natural magnet for the tech industry. It is therefore hardly surprising that the IT industry took root in this city. Major public sector undertakings like Bharat Electronics Limited (BEL), Indian Telephone Industries (ITI) and Hindustan Aeronautics Limited (HAL) as well as a host of government research organizations in the fields of defence and space had already established operations in (then) Bangalore in the 1970s. Therefore, it was a natural choice for the US-based company Texas Instruments to set up its development centre in this city in 1985, when it decided to come to India—with great foresight—to tap into the country's engineering talent. Slowly, an increasing number of IT software companies began to establish themselves in Bengaluru, including Infosys and Wipro (both amongst the top four Indian IT companies today). The momentum accelerated, and by the late 1990s, Bengaluru had become the epicentre of India's IT industry.

Bengaluru undoubtedly had certain historical and natural advantages that facilitated its emergence as a tech centre. However, the government of the state (Karnataka) too had a big hand in this. At a critical juncture, the then chief minister

(Mr S.M. Krishna), assisted by a few key aides and the bureaucracy, played a vital role. They did this through industry-friendly policies, efficiently executed, and a proactive approach towards solving problems and stimulating growth.

It was in this setting that a new player entered the fray and sought to grab the limelight. Clearly, given the dominance of Bengaluru, this was not going to be easy. Yet, N. Chandrababu Naidu, who became chief minister of the neighbouring state of Andhra Pradesh in 1994, took on this challenge. Functioning like a corporate chief, he soon became known as the CEO (rather than chief minister) of Andhra Pradesh, a title he rather fancied. He mesmerized the CEOs of Indian and multinational corporations by personally making PowerPoint presentations, selling his state (and particularly its capital, Hyderabad) as the ideal destination for IT companies. His marketing skills were complemented by his execution abilities: here was a chief minister who 'walked the talk', with marketing promises converted to reality on the ground in a short time. Thus, allocation of land, provision of ready-to-use offices, uninterrupted power, various approvals, all were ensured at double-quick speed.

Till the late 1990s, it was no easy task for corporate CEOs to get an appointment to meet a minister, leave alone meeting the chief minister. Intermediaries or contacts had to be used and even then the appointment was bestowed as a favour. CEOs—especially foreigners unfamiliar with India—had to be forewarned about what to expect: a slow process of getting an entry pass, a long wait in an overcrowded anteroom, an inevitably delayed—sometimes even cancelled—appointment and finally, a meeting with various distractions (phone calls, other visitors, a TV screen and, as a special favour, tea service).

In this long-standing scenario, Mr Naidu's approach was not just a breath of fresh air, but a gale-storm that blew people off their feet. Here was a chief minister who sought appointments with CEOs, visited them in their offices and went to them with ready responses to their 'asks': a role reversal that was, indeed, difficult to believe. It was this innovative strategy—backed by concrete actions on the ground—that catapulted Hyderabad from a non-entity to being at the top of companies' 'consideration set' for places to invest in. In but a few years, Hyderabad became a major IT hub for both Indian companies and MNCs.

The personal hard sell was backed by much work on the ground. Mr Naidu realized that one of the constraints faced by Bengaluru—and a point of frequent complaint—was its physical infrastructure. Bengaluru's very success had become its Achilles heel: its rapid growth and greater prosperity meant that its roads could not cope with the burgeoning traffic, with ever more cars being added daily. Nor could its airport handle the growing passenger traffic, and the new airport was taking far too long to come up. This, Mr Naidu decided, was an area in which Hyderabad would out-compete Bengaluru. Hyderabad too had an old and inadequate airport (shared, as in Bengaluru, with the Indian Air Force), but it managed to get its new airport up and running before the one in Bengaluru. Equally importantly, work on the road connecting the airport to the major IT hub in the city was expedited. An elevated highway from the airport to the city centre was planned and work on it begun. The widely shared perception was that Hyderabad would soon have far better infrastructure than Bengaluru.

During this period (the late 1990s to 2004), few contested

the view that Hyderabad was the clear winner in the investor-perception sweepstakes. From being a near non-entity as far as private investors were concerned, particularly in the booming tech sector, it became the front runner. The catalyst for this change was the innovative strategy adopted by Mr Naidu: an example of how a no-cost innovation can result in big and concrete gains.

Hyderabad exemplified the wonders of smart marketing. Yet, there also needed to be strong underlying basics: packaging is important, but the product needs to be good. In the case of Hyderabad, there was not just the promise, but also the actual visibility of better infrastructure; in addition, there was the vital advantage of availability of a huge talent pool.

In the decade from 1995, Chennai (and the state of Tamil Nadu) too was an active competitor for investment. It was not as flamboyant as Hyderabad, and not as naturally attractive as Bengaluru; yet, its quiet efficiency and vast pool of technology talent soon made Chennai a greatly favoured destination. Being a seaport gave it a unique advantage over its other two southern cousins, helping it to attract investments in manufacturing as well (especially in the automobile sector).

The elections in 2004 saw the parties led by Mr Naidu and Mr Krishna fare unexpectedly poorly in the state elections, with both losing their chief ministerial position. The political establishment attributed their defeat to excessive emphasis on the IT sector, which was painted as being urban, elitist and focused on foreign countries (as opposed to the domestic market). As a result of this, political attention moved back to the rural-agriculture-populist domain. Competition amongst states and innovative policies to stay ahead of others moved to the back-burner.

Fortunately, more states were becoming aware of the benefits flowing from investments by IT companies. As a result, competition amongst states, though muted, did not quite die out. In fact, the number of states actively competing to attract investment increased over time.

In 2014, the erstwhile state of Andhra Pradesh was bifurcated into Telangana and a new (truncated) Andhra Pradesh. Elections followed, bringing Mr Naidu back at the helm of Andhra Pradesh. He announced ambitious plans for his state, including a new capital city and a start-up hub in Visakhapatnam. The boom city of Hyderabad is now part of Telangana and the new government there has also announced big plans for development—of the state, as also of the city of Hyderabad.

Earlier, in 2007, a huge agitation (initiated by Mamata Banerjee, who later led her party to a big victory in the West Bengal state elections in 2011, and became chief minister) resulted in problems in land acquisition in West Bengal for the Nano car project of Tata Motors. With extraordinary speed and much foresight, Gujarat promptly stepped in. The chief minister at the time, Narendra Modi, is reported to have immediately called up Ratan Tata, then chairman of the Tata Group, with an offer of land in Gujarat. In a matter of days, the deal was done, and Tata's production facility for Nano now operates from Sanand, near Ahmedabad. Another sign that competition amongst states did not die out and was back in vogue.

In many ways, Gujarat, under Mr Modi's leadership, emulated the Naidu/Andhra formula: proactive and astute marketing, underpinned by a focus on ensuring superior infrastructure, efficient and helpful bureaucracy, and quick decision-making. Investor meets, called Vibrant Gujarat, were very well organized and considerable effort went into ensuring the presence of the biggest industrialists and investors. An industry-friendly policy, implemented with minimal red tape, and efficient execution of key infrastructure projects: these were the key elements that made Gujarat attractive for investors. The perceived success of the 'Gujarat Model' was amongst the factors that contributed to the big win by the Bharatiya Janata Party (BJP) in the 2014 general elections. Most observers credit Mr Modi, personally, for the BJP's large win. Now that he is prime minister of India, people's expectations are sky-high. However, translating and scaling the Gujarat Model into one for the whole, vast and diverse country has considerable challenges.

Maharashtra has also been one of the leading states in India as far as the IT industry is concerned, with Mumbai and Pune being major centres. It is also a front runner in several other sectors as well, such as engineering, automobiles and media (especially film). Its competitiveness is due largely to a 'first mover' effect, with new companies being attracted to places where their peers are, as also to the availability of talent and of reliable electric power (particularly in Mumbai). Many would argue, with reasonable justification, that the state government itself has done little by way of proactive steps to attract investment or to improve the overburdened infrastructure, especially roads and housing. In fact, the government has sometimes come across as being almost smug, with an 'investors will come anyway' attitude. In late 2014, a new government came to power in the state, bringing with it the hope that it would be far more proactive and dynamic in enhancing the state's competitiveness.

Most of the states not already discussed here have serious constraints in being competitive and becoming hubs of entrepreneurship and innovation. Having and attracting the required human resource base is a major obstacle; the lack of a 'technology temper', industry base and infrastructure (both hard and soft) are other limitations.

Meanwhile, competition amongst states is back on the cards and this spirit is likely to be most visible in the erstwhile Andhra Pradesh, now split into Telengana and the residuary Andhra Pradesh (the latter, as noted, once again led by Chandrababu Naidu). While both began by apparently competing in populism, announcing the waiver of agricultural loans, efforts are also being made to attract investment. As can be expected, Andhra Pradesh has been proactively facilitating investments and wooing investors and entrepreneurs. Not to be left behind, Telangana announced a new industrial policy on its first anniversary (June 2015), which includes an innovative (at least in terms of branding) 'Right to Clearance' for industrial projects. For every day of delay in clearance, the state will fine the official concerned Rs 1,000.

To forestall lost opportunities, Tamil Nadu and Karnataka will have to join the fray. The former continues to take low-key but positive steps to improve the entrepreneurial environment in the state. The 'Tamil Nadu Vision 2023' document says the state aims: 'To be one of the top three preferred investment destinations in Asia and most preferred within India with a reputation for efficiency and competitiveness.'

Recognizing the role of innovation in this, it also mentions its goal: 'To become [an] innovation hub and knowledge capital of India.'

In 2014, Tamil Nadu entered into a partnership with

the Abdul Latif Jameel Poverty Action Lab (J-PAL) to institutionalize the use of evidence in policy-making by rigourously evaluating innovative programmes before they are scaled up, strengthening monitoring systems and enhancing the officials' capacity to generate and use data. In perhaps a first for any state government in India, the Tamil Nadu government also set up an Innovation Fund, with an annual allocation of Rs 150 crore, through which any government agency could access resources for pilot innovation programmes through a competitive process.[1]

Competition is back in vogue. Once again, innovations in policies, approach and procedures will be necessary. One hopes that such positive competition will spread to other states, where the competitive spirit has been lacking in the past.

In this, Rajasthan has been the first off the block in one area of importance, seen by many as one of the impediments to business growth—labour law reforms. The state has announced its intention to bring about substantial changes in the labour laws, and is sure that this is going to be a competitive advantage in attracting investment to Rajasthan. Already, Madhya Pradesh and Maharashtra have followed suit, announcing changes in labour laws. With increasing one-upmanship, based on proactive state policy, one might say that India is entering a new era of 'statesmanship'!

In the years to come, it is entirely possible that such competition will extend to the social sphere. States competing to better each other in health and education facilities, and to outperform on indices like infant and maternal mortality, literacy, quality of education, etc., would be ideal.

---

1. From a report in *The Indian Express*, Delhi, 7 February 2015.

The focus so far has been on competing to attract investment. To do this, states have concentrated on putting in place business-friendly policies, better infrastructure, efficient execution and simplified procedures. These factors continue to be important; but increasingly, innovation is going to be the magnet for investments. States that are perceived as centres of innovation will see investment flowing in far more easily than others. Also, at the micro level, innovative products and ideas will attract angel- and venture-funding in the early stages, and larger private equity funds as success manifests itself. This will encourage other entrepreneurs, setting in motion a virtuous cycle of innovation, entrepreneurship and investment. Thus, innovation is going to be the driver of both competitiveness and investment. In this emerging world, states that top in innovation are going to be the winners.

In this context, what can states do to enhance their attractiveness?

States can take a number of initiatives to promote innovation and become centres for creativity, innovation and entrepreneurship. This is important not only to attract investment, but also to increase productivity—resulting in the creation of jobs, economic growth and social benefits. Countries—and the states and cities within them—have realized the role of innovation, and are now vying with each other in this sphere. It is in this competitive global context, and not merely intra-country competition, that states in India will need to create policies and take steps to facilitate innovation and attract innovators.

What each state will have to do depends on its specific context (including history, culture, geography and human resources) and its socio-economic position. Broadly, at the level of the state, the factors that determine innovativeness and which the state may therefore seek to address include the following:

1. The socio-cultural values regarding entrepreneurship, risk-taking and failure. Often, these are deeply embedded and are based on centuries of tradition (for example, communities that have long been in employee—as against business—roles). Changing such long-held values is not easy and is generally a slow process. However, steps to hasten and accelerate the change are possible. These include:

   • Ensuring the easy availability of funding for start-ups—through grants and soft loans by the state and by providing fiscal incentives to angel investors; providing other facilitation to entrepreneurs (ease of acquiring land and at reasonable prices, water and power connections, ready-to-use space, etc.); minimizing the requirement of clearances/approvals.

   • Giving special recognition to entrepreneurs and showcasing their work; promoting role models of successful entrepreneurs, so that youngsters are inspired to emulate them.

   • Creating start-up fellowships so as to ensure some guaranteed income (for, say, a year) to those wanting to take the risk of becoming entrepreneurs.

   • Introducing innovation fellowships at the school and college level, as a means of encouraging and nurturing innovativeness.

   • Giving special recognition to 'grand failures': those who came up with an exciting innovative idea, which unfortunately did not finally work out.

2. Supporting and encouraging diversity in all aspects is an important way for the state to create an innovation-friendly environment. This means getting away from a parochial mindset and adopting a cosmopolitan outlook. The former is often a shortcut to political power, so such a transition will not be easy, but the payoff in terms of economic growth and job creation (through innovation) should be equally politically attractive, especially in today's world. It is necessary for governments to move away from special reservations for 'sons of the soil' and excessive enforcement of the local language (or cuisine or movies, as is being proposed in at least one state in the country). Colleges, for example, could have seats reserved for toppers from other states. Equally important is a political climate that permits, even encourages, diversity of thought and idea. Dissent is an essential catalyst of innovation.

3. Providing a strong support system for innovation by creating the necessary infrastructure—test centres, rapid prototyping facilities, Wi-Fi hotspots for Internet access, co-working centres and incubation facilities. These need not be set up by the state, but it can provide the encouragement (including loans or grants) to facilitate this. In addition, innovation parks or zones where industry, academic and research and development institutions and incubation centres are co-located could be created.

4. Ensuring the support infrastructure: not only good physical and electronic connectivity through roads/ highways, rapid transit systems, air/rail links, high-speed, reliable and cheap electronic connectivity and widespread

Wi-Fi, but also social infrastructure like schools, hospitals, restaurants, shopping centres, theatres, etc.

5. To promote innovation in the social/developmental arena, it would be worthwhile for the state to consider setting aside a part of the budget for each scheme or project specifically for promoting innovation within that project. This could serve as an award or challenge grant given to the most innovative idea(s) that could help the project (or its intended beneficiaries) through greater efficiency or effectiveness. At a more general level, the state could fund 'grand challenge' prizes (see Chapter 7) for innovative solutions to a few of its biggest problems.

There are other specific and general actions that a state could take, depending upon its strengths, opportunities and challenges. The purpose must be to attract innovators, create innovators, facilitate innovation and put in place an ecosystem that is conducive to all three.

How does one compare states with regard to innovation? Are cities, in fact, a better unit for comparative assessment? What factors drive competitiveness and innovation at the city and state levels? We will discuss these issues in what follows.

Many of the factors that account for the competitiveness and attractiveness of a location for investment are determined by the policies, attitude and efficiency of the government of the state concerned. However, some of the determinants are more localized and are city-specific. This is all the more so for innovation, in which local factors play an even bigger role. Before looking at how different cities in India stack up, it is worthwhile to examine the factors that influence the innovativeness of a city and the relative advantages or drawbacks of some of the cities.

The innovativeness of a country, like its competitiveness, is built up from local to regional to national: from city to state or region to country. In the case of competitiveness, many of the factors are sector-specific. For example, the transportation infrastructure (roads or rail connection to ports, efficiency of ports) is a vital element of competitiveness for the export of manufactured goods, but is hardly as critical in the case of export of IT software and services.

Innovation, unlike competitiveness, is typically driven by factors that are less sector-specific. This is not to say that all the determinants of innovation are sector-agnostic, but typically entities or regions that are innovative in one field tend to be innovative in other fields too. Of course, over a period of time, the organization or area may become well-known for its innovation in one field, but a closer look will show that it is as adept in other areas too. Today, Silicon Valley is synonymous with innovations in information technology. However, it is also home to cutting-edge start-ups in renewable energy. Further, California has been the hub for many breakthroughs in electronics (Hewlett Packard was amongst the earliest 'garage start-up' companies). Long before the Silicon Valley was christened, San Francisco's Bay area was innovating with lifestyles and gave birth to the hippie culture, amongst other things. In the 1960s, it was also the hub of the anti-(Vietnam) war demonstrations in the US and the peace movement (with its 'Make Love Not War' slogan). Clearly, the area is a centre of innovation, across sectors of industry and spheres of life.

What is true of a geographical area appears to carry across to organizations too. For example, Google is not only about innovative search algorithms; equally, it is about a creative business model (with no payment by the user and beneficiary) and great marketing. It is also about innovations in other areas, including driverless cars and Google Glass. Like California, its innovations too are not limited to one field; innovation is inherent to, and embedded in, the organization. Thus, innovation seems to be linked to the area or organization: its culture, demography, ambience and policies.

Innovation, ultimately, is a product of people: often an individual, sometimes a team. However, there are external factors that can either catalyse and drive this, or impede it. One crucial element is socio-cultural: what is the societal attitude to risk? How is failure perceived and dealt with? What is the attitude towards business? These aspects change with time, as is clearly exemplified by the metamorphosis in India. Till the 1980s, business in India was looked at with suspicion by not only the government, but equally by most people in the country. The image of a businessman was generally that of a wheeler-dealer, getting his way by bribing officials and making profits by selling poor quality goods (or services) in a self-created monopolistic situation. Organized business was seen as the preserve of a small class of business families and, therefore, hereditary. Small businesses and their owners received little social respect; big businessmen were held in awe because of their money power, but they too generally evoked little social respect.

From the 1990s, this changed radically. As knowledge rather than money became the real capital for business in new areas like IT, a new breed of businessmen emerged. They were, in general, not from traditional business families and were well-educated. They brought into their businesses the traditional middle-class values of thrift, integrity, honesty and hard work. Soon, they became the icons of a new India. This new breed of techno-entrepreneurs is best exemplified by N.R. Narayana Murthy, founder of Infosys. At that time, in the 1990s, practically all middle-class Indians aspired for a 'regular' job, preferably one with assured career-long employment (with a government job being amongst the most desirable). Pressure from parents, relatives and peers served to drive youngsters in this direction. Employment in the private sector was generally perceived to be risky and uncertain. Slowly, from the 1980s, jobs in the private sector began to look attractive, as the salary differential—especially for managerial and professional positions—between the private and public sectors grew. As a result, employment—as compared to business—became even more attractive.

The IT boom and the success of start-ups (most visibly, Infosys), as also economic reforms, changed the social outlook in the 1990s. Entrepreneurship, profits and even business were no longer frowned upon. Infosys and Murthy were widely cited as an example of the fact that business could now be done in a 'clean' manner and that pay-offs, corruption and bribes were no longer integral parts of the business environment. The rise and outstanding success of Murthy, a person from an ordinary middle-class family, provided hope and inspiration to millions of youngsters. The aspiration changed from employment and job security to dreams—as in the West—of a start-up and an IPO (an initial public offer of shares on the stock market). With stories of hugely successful start-ups from the West and, increasingly, from within India—widely highlighted by media—there has been an accelerating aspirational desire to become an entrepreneur. This trend is particularly pronounced amongst graduates of engineering and management schools, but is even visible in fields as diverse as fashion design and finance.

As a result, an increasing number of young persons are now focused on entrepreneurship. A fair number become entrepreneurs immediately after graduation; many others gladly take up jobs in start-up ventures, despite the risk involved. Of those who take up employment in large and well-established organizations, a significant proportion do so to gain experience (and/or to save some money) before embarking on an entrepreneurial venture. The comparative ease of getting seed- or angel-funding, the lower need for capital and the availability of incubation facilities (often in their own education institution) has encouraged young graduates—sometimes even before they graduate—to follow an entrepreneurial path. Further impetus has come from new and younger role models—replacing Murthy and his contemporaries of yesteryear—who have made it big: the various Bansals from the e-commerce world, Deepinder Goyal of Zomato, Bhavish Agarwal of Ola, and a host of others.

Entrepreneurship is, of course, not synonymous with innovation. However, in sectors with established and strong incumbents, a start-up company has necessarily to be innovative to break into the market in most cases. On the other hand, foraying into a new field is obviously being innovative. Thus, innovation is related to entrepreneurship and the extent of the latter depends on societal attitudes towards businessmen and risk. When one looks at the innovation potential of a city it is therefore necessary to assess the societal norms. There is no standard index or quantitative measure for this and so it must be qualitative, based on anecdote and judgement.

Another major socio-cultural factor is the attitude towards wealth. Till the 1990s, much of Indian society considered many other attributes to be at least as important as money. Thus, even if wealth was not looked down upon, things like education, skill, knowledge, culture and spirituality were considered very important. The teacher and priest were greatly respected; amongst the middle-class, a degree (certifying that one was 'educated') was important and, with obvious gender bias, proficiency in cooking, embroidery, classical music and dance were all signs of accomplishment and achievement for girls. In due course, education became important for girls too. Today, India is becoming increasingly materialistic. People tend to be judged, even respected, based on how much (money) they make, rather than how much they know. This change, which many date to the economic transition in 1991, has given a huge boost to entrepreneurial dreams; for even a successful employee can generally earn only a fraction of what a thriving businessman does.

Like any other socio-cultural variable, this 'love of money' index is difficult to quantify. Again, one can best capture it only through observation, understanding of context and judgement. As in the case of the attitude towards business, discussed earlier, entrepreneurship triggered by greed also drives innovation. After all, the quickest way to make big money is to innovate a new product, service or business model.

Just as entrepreneurship and innovation are linked, the two socio-cultural drivers discussed above (attitudes towards risk and wealth) are also linked. If 'greed [...] is good' (as Gordon Gekko declared in the Hollywood movie *Wall Street*), then businesses and businessmen who make money must be all right too. Gordon Gekko also added, '[...] greed is right. Greed works. Greed, in all its forms […] has marked the upward surge of mankind […].' In reverse, if businessmen are respected, then greed is fine. Therefore, when one looks at a society—or a city—while these two factors can be separately gauged, they will probably be in tandem with each other.

In addition to these two socio-cultural factors, there is a third external factor that is, arguably, a big driver of innovation: adversity. Faced with a serious, near-insurmountable problem, the brain searches for solutions. When none is easily found, it looks for different or creative solutions—the out-of-the-box thinking that gives rise to innovation. This is the reason why one sees so much innovation at the grass-roots level in India; for life at that level—whether in cities or in rural areas (through probably more so in the latter)—is full of adversity. Many of the solutions are not necessarily breakthrough innovations, but are more in the nature of improvisation (jugaad). Yet, this out-of-the-box thinking could culminate in a major innovation.

Adversity is not limited to those at the lower end of the economic spectrum. Middle-class living also throws up a fair amount of adversity in India. Power outages, water shortages, traffic jams and bureaucratic hurdles are common events in the life of a middle-class urban Indian. These adversities have each been met by crafting a solution or a workaround approach. Sadly, many of the solutions are selfish or anti-social. Thus, water shortages are overcome by putting a booster motor in the pipeline—even though, as each household does this, it soon becomes a pointless exercise. Alternatively, water is pumped out from underground bore wells. As everyone does this, each has to go ever deeper. Of course, both solutions are illegal, and ultimately self-defeating. The only positive feature is the mindset of trying to find alternative, and innovative, approaches. The challenge, discussed later in greater detail, is how to channelize this innate mindset (probably embedded through centuries of living in adverse circumstances) to productive innovation.

The first two external factors mentioned—the attitudinal changes towards business/entrepreneurs and towards money or wealth—are primarily of importance amongst the middle class. The third (adversity) is certainly a much bigger factor for the poor, but has its impact on the middle class too, as discussed above. Adversity or challenges faced in daily life are thus a catalyst and also a possible indicator of innovation. Measuring economic adversity is obviously easy, but assessing the extent of adversity in day-to-day living is a more difficult and complex exercise. Hence, quantification of these drivers of innovation is not easy.

While adversity has historically been an important trigger for innovation, one could also argue that its antithesis—prosperity—can be equally important. After all, prosperity gives the individual, organization, society or country the resource-cushion that permits risk-taking. The willingness to take risks comes when one is so poor that there is nothing to lose or—at the other extreme—has so much that one can afford to risk a part of one's resources. In the latter case, many would look at it as an investment—for example, in high-risk research and development projects.

Diversity is another, and major, driver of innovation. At the societal level, a diverse population means that each individual is used to the idea that those around him or her speak a different language, look different, follow a different religion, cook dishes that are different and have a different experiential base. The diversity could be of gender, ethnicity, religion, race, caste, economic class, education level or age. Given this, persons who grow up in a diverse milieu naturally accept the proposition that others may think differently, too. There is thus an easy acceptance of diversity of opinion and thought. Such diversity is clearly the foundation of innovation, which has its genesis in thoughts and ideas that differ from the conventional.

The freedom to think differently, to voice and discuss an unconventional idea and take it to fruition through a product or service: this is clearly an essential element of any milieu that seeks to encourage innovation. Be it a classroom, community, organization or country, the degree of freedom to dissent is thus amongst the indicators that may predict innovativeness. True democracies are therefore likely to be more innovative. This is not to say that we will not see innovation in countries with totalitarian governments; however, it is likely that they might be even more innovative if they have greater freedom. There are nuances too: some countries may be politically totalitarian, but allow a great deal of freedom in other spheres—cultural, economic or technological—in which they may therefore be very creative. What is noted briefly about countries is equally true about organizations, but more on that later. As for cities, their general milieu will obviously be determined at the larger level (country or possibly state), but there is arguably also a certain 'culture' in each city, determined by social mores and norms, which is either constraining or liberating.

Determining which locations are the most innovative, or have the greatest potential to be centres of innovation, has to be based on a combination of data and judgement. In India, reliable quantitative data on many relevant factors is generally not available and therefore an assessment has to rely considerably on qualitative analysis. Also, as discussed earlier, the city is a more appropriate entity for assessment rather than a state, and city-wise figures are even more difficult to come by. Therefore, rating cities in terms of their present or potential innovativeness must be based substantially on an analysis of various qualitative factors, many of which have been discussed earlier in this and other chapters. Amongst the factors that need to be taken into account are:

1. The availability of appropriate talent in sufficient numbers and the ability of the city to attract such talent from elsewhere. The latter must include multinational or expatriate talent. Given the growing importance of technology as a source and catalyst of innovation, the talent pool must also include technologists.

2. Attracting and retaining youngsters is vital to ensure a large talent pool and a necessary condition for innovation; therefore, the needs of this talent pool are paramount. These include safety, clean air and water, entertainment (restaurants and movie theatres), health facilities, shopping centres, gymnasiums, high-quality schools, good housing, easy mobility and, importantly, an outsider-friendly social environment. These elements define the quality of life in a city and are important determinants of attractiveness.

3. Easy availability of risk-funding, along with mentorship. The former could come from governmental institutions or banks, but mentorship requires individuals with experience in entrepreneurship and its challenges.

4. Infrastructure, especially of a type that is conducive to innovation. This means not just low-cost workspaces with facilities like high-speed Internet connectivity, but co-working or incubation centres, which promote

interaction amongst entrepreneurs. The city itself must be well connected to the outside world.

5. A strong support system of related facilities—for example, rapid prototyping centres, test facilities, consultants or academic experts in specific fields, beta (test) sites.

6. A good base of relevant institutions. These could be educational institutes, research laboratories, and industrial companies in the particular field.

7. Supportive policies of the state in which the city is located. Special incentives for entrepreneurship/start-up companies and for micro, small and medium enterprises (MSMEs); incubation centres; tax concessions or lower rates for power and real estate; training facilities or incentives for training/employment: these are specific steps that encourage innovation.

8. The intangible and not easily definable social ethos of the city is another important factor. This includes its 'technological temper', cosmopolitanism and acceptance of outsiders, positive attitude to risk-taking, openness to change and new ideas, diligence or work culture. Many of these are determined by history and the evolution of the city's social milieu. In exceptional cases, government policy (given other supportive factors) can help create, through scale and speed, a new social environment. A case in point is Gurgaon: the scale and speed of its growth (thanks to conducive government policies) was such that its new culture—brought in mostly by young professional outsiders—swamped the old through sheer numbers of immigrants. It became a new city with a new culture.

Of the above factors, some can change quickly (government policy, for example) and some slowly (availability of social infrastructure like schools, hospitals and shopping centres). A few are givens (history) or evolve over long periods (culture).

Therefore, depending upon the extent of pro-innovativeness of each factor, the ranking of a city can change quickly. For example, if other factors are positive and policies are the major hindrance, a positive new policy may overnight catapult a city much higher on the innovativeness ranking. This, and infrastructure-related factors, provide leverage to the government to take proactive steps to increase the innovation potential of a city.

On the basis of the elements listed above, it should be possible to collect primary data, collate secondary (available) data and make a qualitative assessment of non-quantifiable factors to evolve an index of innovativeness. Doing so is beyond the scope of this work, but it is an effort that an academic institution or a research agency could well take up. One hopes that state governments will fund and support such an effort (on a regular, say yearly, basis) in order to promote their most innovative city. The central government could also do so, as a means of promoting innovation by creating competition amongst cities to top the ranking. This could spur positive action at the city and state levels.

As one example of the kind of indirect data that one could look at, consider the following. In October 2014, the Kendriya Vidyalaya Sangathan (KVS, which controls a large number of schools across India that are affiliated to the CBSE decided to discontinue offering German as an alternative to Sanskrit, which was part of the 'third language' requirement. The decision affected more than 65,000 students then studying

German in over 500 KVS schools. A subsequent survey[2] by *The Times of India* and IPSOS (a French market research agency) of students and parents in eight cities provides some interesting insights into what one might call the 'outward orientation' of these cities. Despite the constraints of such a (limited) survey, the vast range in the responses is amazing. The most telling one is the response to the question of choice between a foreign language (not necessarily German) and an Indian one as the third language. In Ahmedabad, only 18 per cent of the students and 15 per cent of the parents preferred a foreign language; Hyderabad too had only 23 per cent and 22 per cent preferring a foreign language; for Chennai, the corresponding figures were 100 per cent and 82 per cent, and in Pune, they were 82 per cent and 96 per cent respectively. Unlike these four cities, Bengaluru—with 43 per cent and 10 per cent—seemed to have a wide generation gap between students and parents.

Lest one interpret this as an indicator of a particular city's interest in engaging with the outside world, another set of figures gives cause for caution. While the foreign language preference amongst (school) students in Hyderabad was low, it was this city that sent the maximum number of students to the US between 2008 and 2012 (26,220; in addition, one should add 2,969 attributed to the twin city of Secunderabad). Mumbai was a distant second, with 17,294 students sent to the US. Other cities (Bengaluru, Chennai, Delhi and Pune) were all below the five-figure mark.[3]

---

2. Reported in *The Times of India* (Delhi), 29 November 2014.

3. Figures are of F-1 (student) visa, as reported in 'US-bound tech students: 4 Indian cities in top 10', *The Times of India* (Delhi), 23 November 2014.

These figures do not directly impact innovation or innovativeness. However, there is a strong case for assuming that an interest in learning a foreign language is linked to a mental framework that is broader, outward-looking and therefore probably more open. A correlation between going abroad to study and innovation is far more tenuous, and it would not be appropriate to assume a linkage.

Another set of interesting figures is from the first edition of the ET Start-up Awards, announced in August 2015. Of the start-ups that were shortlisted for the awards, fifteen were from Bengaluru, twelve from Gurgaon/Delhi (surprisingly, none from Noida), eight from Pune/Mumbai, five from Hyderabad and just one from Chennai. There were none from Kolkata.[4]

These figures too are but indirect indicators of innovation, but probably provide a better correlation to innovativeness and of the ecosystem conducive to it.

Based on the factors discussed above, the next chapter assesses a few cities in greater detail, focusing mainly on qualitative considerations, and ranks them in terms of innovation.

4. This break-up of the shortlist for the awards is from 'Salute Independence, Freedom to Enrich', editorial from *The Economic Times* (Delhi), 15 August 2015.

6

INDIA'S MOST INNOVATIVE CITIES

Indian cities: historical factors that affect innovativeness; present state and potential. Ranking the nine most innovative cities in India.

Based on the factors discussed and outlined in the previous chapter, we can assess a few cities in India and look at their comparative advantages. As noted earlier, most, if not all, of these factors do not easily lend themselves to quantification or objective measurement. Such an assessment is, therefore, necessarily qualitative and subjective.

Bengaluru has, over the last two decades, won increasing recognition, nationally and globally, as the hub of the Indian IT industry. In fact, its fame has earned it mention in US presidential speeches. In some quarters in the US, it has been converted into a verb: to be 'bangalored' is to lose your job (in the US) because it is presumably outsourced to a company in Bengaluru! The city is not only the centre-piece of the IT industry, but has also long been the nucleus of the aerospace sector.

Beginning with the establishment of the government-owned Hindustan Aeronautics Limited in 1964 in the city, Bengaluru has been the country's biggest base for the aircraft industry. Institutions like National Aeronautics Laboratory and a number of other research laboratories, along with the academic prowess of the Indian Institute of Science (one of the few Indian institutions featured in global rankings of the top 500 universities) have ensured Bengaluru's domination in this area. Several facilities of Indian Space Research Organisation (ISRO) were set up from the 1970s onwards which added the space dimension, making Bengaluru the aerospace capital of the country.

In days gone by, companies like Indian Telephone Industries (ITI), BEL and Hindustan Machine Tools (HMT) were household names. While ITI and HMT are now hardly in the pink of health (both may well shut down), in their heyday they helped to add substantially to the technological prowess of Bangalore, where their major facilities were located.

All these institutions attracted engineering talent from around the country. There is no hard data available, but this almost certainly made Bengaluru the city with the highest number of engineers in the country. The big and well-known companies and institutions offered attractive jobs, which encouraged talent to flow into Bengaluru. However, geographical and cultural factors played a role too. The city's location made it accessible reasonably easily and quickly from other parts of India. Earlier, when people travelled almost exclusively by bus or train, this was not unimportant. In more recent years, when air travel has become more common, access is much easier and faster.

Geography bestows another advantage: the city is blessed with a salubrious climate, thanks to its location and altitude. It never gets too hot or too cold, and the monsoon rains are generous but not overwhelming. Its location also puts it within quick reach from major cities in South India and from Mumbai and Pune, making it easy (and relatively inexpensive) for prospective employees from these cities to travel back home from Bengaluru.

It was probably the location and climate that prompted the British, in the colonial days of the Raj, to establish a military cantonment in the city. The spin-offs from the cantonment (visible, too, in other locations where they were set up) included English-medium schools, widespread use of English, greater Westernization and modernization and a more cosmopolitan outlook. These cultural effects make the city relatively 'outsider-friendly', helping to attract talent from around the country and, increasingly, from around the world. Expatriates find that the culture of the city facilitates their transition and makes it an easy place to relocate from their home country.

As a result of all this, the confluence of history (the British presence), geography (climate and location) and policy (the setting-up of technological organizations) has made Bengaluru a magnet for talent. New technologies and new thinking have triggered entrepreneurial dreams, particularly amongst Bengaluru's huge technological talent pool. Start-up companies based on innovative new ideas are the result. A conducive ecosystem has facilitated the process and, as experienced elsewhere, the growth of entrepreneurship rapidly snowballs, as start-ups feed off each other and create a positive feedback loop for ecosystem development. Today, Bengaluru is the hub of start-ups, and leads in innovative proposals for new ventures. While many of these are in the IT field, given Bengaluru's domination of that space, the city is also home to a growing number of start-up companies in fields like health, medical electronics, biotechnology and aerospace. Of late, the rapid growth in mobile communication has spawned a number of innovative ventures in mobile applications and telecommunications.

With new-generation start-ups in a wide range of technological areas, Mumbai seems to be once again demonstrating the kind of innovation and entrepreneurial spirit that led the first wave of IT in the country. It suffers from inadequate infrastructure, high costs, many years of average (at best) governance, traffic jams and long commute times and a whole host of other woes. Yet, the sheer energy of the city, its professional and businesslike culture, its huge talent pool and its global linkages make it an attractive base for innovative new ventures.

Mumbai does not have a well-known or celebrated history, unlike Delhi, for example. The city (formerly Bombay) comprised seven distinct islands, which in ancient times were part of the kingdom of Emperor Ashoka, and later of the Silahara dynasty. The rulers of Gujarat annexed the islands in 1343.[1] A Portuguese attempt to conquer Mahim (one of the seven islands) failed in 1507, but in 1534 Sultan Bahadur Shah, the ruler of Gujarat, ceded the islands to the Portuguese. In 1661, the Portuguese handed over the islands to the British, as 'dowry' for the marriage of King Charles II and Catherine of Braganza, sister of the king of Portugal.[2] In 1668, the Crown persuaded the East India Company to rent the islands for 10 pounds a year.[3] At that time, the Mughals, the Marathas and the Gujarat princes were all more powerful than the East India

---

1. 'Bombay: History of a City', Learning Trading Places, British Library.

2. Mumbai, *Encyclopedia Britannica*, 2016.

3. 'Bombay: History of a City', Learning Trading Places, British Library.

Company. Even British naval power was not at par with that of the Portuguese, Dutch, Mughals or Marathas. However, the decay of Mughal power, the Mughal–Maratha battles and instability in Gujarat drove artisans and merchants to the islands for refuge, and Bombay began to grow.[4] Records indicate that in just seven years (from 1668 to 1675), the population of the city rose from 10,000 to 60,000.[5] As a consequence of its growth, the East India Company officially transferred their headquarters from Surat to the new city called Bombay.

The industrialization of Bombay began in 1857, with the setting up of the first spinning and weaving mill. As a result, by 1860 the city had become the largest cotton market in India. The halt of cotton supplies to Britain due to the American Civil War (1861–1865) created a great trade boom in Bombay. However, cotton prices crashed with the end of the Civil War. Nonetheless, the boom had facilitated the opening of the hinterland, and Bombay became a major centre of trade. The Suez Canal, opened in 1869, gave a big boost to trade with Britain and Europe, and resulted in growing prosperity for Bombay.[6]

Innovation seems to be embedded in Bombay itself. The ambitious projects to link the islands and the reclamation work are themselves innovative and imaginative efforts that have resulted in the creation of a major global city. The Hornby

4. Mumbai, *Encyclopedia Britannica*, 2016.

5. 'The Urban Imagination, Bombay: The Gateway to India', Harvard University (Source: http://dighist.fas.harvard.edu/courses/2015/HUM54/exhibits/show/hornby-vellard/the-islands-come-together--the).

6. Mumbai, *Encyclopedia Britannica*, 2016.

Vellard, linking two of the seven islands, was one of the first engineering projects to be undertaken in Mumbai. This was initiated by William Hornby, the governor of Bombay, in 1782, to prevent the flooding of low-lying areas. It was followed by a series of major civil-engineering projects involving the construction of a number of causeways. As a result, the seven islands were finally merged into one single mass in 1845.[7] Bombay received a further boost in 1853, when the country's first railway connection for passengers was inaugurated between Bombay's Bori Bunder (later renamed as Victoria Terminus, and now known as Chhatrapati Shivaji Terminus) and Thane. In fact, a few other railways are known to have operated in India prior to 1853, for hauling materials.[8]

In 1918, an ambitious scheme for the construction of a sea-wall in Back Bay to reclaim an area of 1,300 acres (525 hectares) of land was proposed. This was completed only after World War II (1939–1945). It linked Nariman Point to Malabar Point through Netaji Subhas Chandra Bose Road (Marine Drive), the first divided highway of its kind in India.[9]

Mumbai's long commercial history has shaped and defined its business-like, efficient culture, exemplified by the '8.48 super-fast to Churchgate'. To those not familiar with the city (and as elaborated in Chapter 2), the 'super-fast' is a local train which, unlike the 'slow' local that stops at all stations, skips many of the intermediate stations between its originating

---

7. 'The Urban Imagination, Bombay: The Gateway to India', Harvard University (Source: http://dighist.fas.harvard.edu/courses/2015/HUM54/exhibits/show/hornby-vellard/the-islands-come-together--the).

8. Indian Railways Fan Club (http://www.irfca.org/).

9. Mumbai, *Encyclopedia Britannica*, 2016.

point and its terminus. The local trains are famed for their punctuality, and the '8.48' will invariably leave at 8.48 a.m. The local trains are the lifeline of Mumbai and the ever-expanding city needs them for moving people quickly, cheaply and efficiently between work and home. Given the dependence of millions on the local trains, these have become an integral part of the culture of the city. Even outsiders quickly absorb the '8.48 syndrome'. They learn that missing a super-fast may mean riding aboard a slow local, which, even though it may depart only two minutes later, will get them to their destination fifteen minutes (or more) later. The local trains thus help create a culture where punctuality is important and time is valued, thereby contributing to the businesslike efficiency of the city.

Apart from its culture of efficiency, what adds to the city's attractiveness for innovative entrepreneurs is the fact that it is a financial hub and a centre for private equity, venture capital and angel investors. For decades, its feature-film industry (known as Bollywood) attracted large numbers of youngsters with dreams of stardom; now, the city is also a magnet for talented youth with dreams of starting the next billion-dollar company.

Mumbai's near-neighbour, Pune, is another incubator for start-ups. The city has long been a major industrial hub. Like Bengaluru, it has a salubrious climate (though the summer is distinctly warmer), and was a British cantonment, leaving it with a strong legacy of the English language and a more cosmopolitan outlook. Yet, it has very strong and deep indigenous cultural roots (far more than Bengaluru) and has long been regarded as a (if not *the*) centre for Marathi theatre, literature and music. Over the years, Pune has become a major hub for education and possibly boasts the highest proportion of foreign students studying in Indian universities. At some point it was called the 'Oxford of the East', and it rather fancies this tag line. While it has some similarities with Bengaluru, their patterns of early industrialization were quite different: whereas Bengaluru boomed with public-sector enterprises, Pune was home to major foreign and Indian private-sector companies. In both cases, it was probably this technology ambience that attracted the IT industry. Today, Pune too is a large and important IT centre, though not quite in Bengaluru's league. It continues to be home to a large number of manufacturing companies and has emerged as a centre for the automobile industry, with manufacturing plants of Tata Motors, Volkswagen, Fiat, Mercedes Benz and General Motors. It is the home base for one of the world's largest forging companies, Bharat Forge, and a host of other engineering companies, both Indian and foreign.

Pune was—and in many parts still is—a laid-back city. The tradition of an afternoon siesta is very strong: even now, many shops are closed from 1 to 4 p.m. In striking contrast to nearby Mumbai, traditional Pune displays an almost non-business-like attitude. Shopkeepers display their wares to prospective customers with visible reluctance. For example, on being asked to show a particular item, they will first tell you its price (implying its unaffordability to you) without removing it from the shelf. They will then ask you to specify colour, size, etc. and take out one item, rather than showing you a range of items and trying to upsell a more expensive product (as shopkeepers in most other cities will do). Similarly, autorickshaw drivers seem reluctant to go to many destinations, irrespective of being offered a higher payment. One would assume that such an easygoing culture would be unsuitable for a tech industry. Yet, in another example of the typical contrasts and contradictions of India, the siesta lifestyle coexists with a booming 24x7 IT-BPO industry, manned by hard-working youngsters. While the vibrant and energetic new culture has not yet completely vanquished the old, the signs of its doing so are visible in the new, no-afternoon-break shopping centres and the always-available new-age taxi services.

It is certainly this engineering-industry base, combined with the tremendous talent pool provided by the extensive educational network, that makes Pune a hot spot for technology start-ups. Its relative proximity to Mumbai (aided by an excellent expressway) and good air-connectivity to many Indian cities adds to its attractiveness. Arguably, its history in creative arts supplements its technological prowess to make it a natural centre for innovation.

In contrast to Bengaluru, Mumbai and Pune, Delhi does not have a history of being a centre for industry. For many years, it was seen as a 'babu' town because of its hordes of low- and mid-level government functionaries. Its main claim to fame was its long history, dating back many centuries. For over a thousand years, it was the seat of power for successive dynasties ruling large parts of the country. Through the years of Mughal rule, it grew in importance and, but for a brief period, was the capital of the Mughal Empire. The advent of the British Raj saw a change, as the capital shifted to Calcutta. For a few decades the city languished, till the British moved the capital back to Delhi in 1911 and created the city of New Delhi. Yet, despite its political importance, there was not a great deal of industrial growth. Post-Independence too, no large public-sector manufacturing enterprise was set up in Delhi. Faridabad, an adjoining town in the neighbouring state of Haryana, slowly became a centre for engineering industry, but on a scale that was a far cry from what then existed in Bengaluru or Pune. The setting up of the Maruti car factory in Gurgaon (also in Haryana) in the 1970s gave a fillip to industrial growth in the region, and was a catalyst in creating what is now a major hub of the automobile industry. Large investments in infrastructure facilitated the growth of the National Capital Region (NCR, comprising Delhi and its immediate surroundings in the states of Uttar Pradesh, Haryana and Rajasthan). While older towns like Faridabad, Ghaziabad and Meerut grew, the new towns of Noida and Gurgaon saw a real boom, particularly in the new millennium. Gurgaon—with its massive multistorey housing projects—began as a dormitory suburb of Delhi, but soon evolved into a major base for the IT-BPO industry and for the auto industry. The availability of high-quality office space, at costs lower than those in central Delhi, led to many corporates shifting their offices to Gurgaon. Transportation infrastructure—particularly the Delhi–Gurgaon Expressway and the Delhi Metro line to Gurgaon—providing excellent connectivity to Delhi and to the airport, has enhanced the attractiveness of Gurgaon. Noida too has similar facilities, and these have greatly spurred the growth of industry in these two cities. Delhi was but a small centre for the IT industry in the 1990s; now NCR is amongst the country's major IT hubs, thanks mainly to Noida and Gurgaon.

The large investment in infrastructure in NCR, especially on roads and the Metro system, has resulted in excellent connectivity within the region. The privatization and expansion of the Delhi airport has facilitated much better air connectivity to destinations in India and abroad through a truly world-class terminal. Aiding industrial growth has been the easy availability of large numbers of people at all levels of skill and expertise, thanks to the huge catchment area for human resources in north, east and northeast India. With few large centres of industrial growth in this very large and hugely populated swathe of India, Delhi/NCR is a natural magnet for talent from the hinterland. Thus, sheer numbers compensate for the lower proportion of educated people in this area (as compared to south or west India), including in the crucial field of engineering talent. However, Delhi itself is a source for high-quality talent, with Delhi University being an acknowledged leader in the higher education arena. In engineering, IIT Delhi is amongst the premiere institutions in the country. There are also a large number of other engineering institutions in Delhi and nearby cities. The lag in the development of infrastructure for higher education in large parts of north, east and northeast India has led school-leaving students from these areas to seek admission in colleges in Delhi. As a result, there is intense competition. In fact, the cut-off marks (based on marks in the school-leaving examination) have actually hovered around 100 per cent for admission to some of the well-known colleges in Delhi University! Thus, the quality of students (at least, as judged by the criterion of academic performance in the school-leaving examination) tends to be high. Interestingly, the highest demand (based on cut-off percentages for admission) amongst non-engineering fields is for courses in commerce and economics, and not for the one-time-favourite science subjects.

One of the reasons for the explosive growth of Gurgaon (and, to a lesser extent, Noida) was the boom in the BPO industry, especially after 2001. Beginning with a great deal of voice-related ('call centre') work, it quickly evolved to add increasing sophistication and value to the type of work handled. The proportion of pure voice work now done in the BPO industry has vastly decreased. In the last few years, NCR has also become a base for a growing number of start-up ventures in the IT space. While many operate from Gurgaon or Noida, a surprising number are based in Delhi itself. There is now a trend towards concentration of start-up ventures. In Delhi, for example, a recent report indicated that parts of north Delhi are emerging as the new hub for entrepreneurs working in core information technology. This locational preference was attributed to the fact that a large number of colleges of Delhi University are in the vicinity and that real-estate costs are lower than in many other areas of Delhi. Also, the metro provides good connectivity to other parts of the city. As in other major locations, NCR too has seen considerable growth in incubation facilities and co-working spaces, indicative of an active start-up ecosystem, and adding to the concentration factor.

Similarly, in Mumbai, the Powai area is now a hotbed for new tech start-up companies, thanks to attractive (by Mumbai standards) real-estate costs, and most importantly, the proximity of the IIT. IIT Bombay has incubation facilities within its campus, and as elsewhere, the presence of some start-ups is itself a magnet for others. This in turn has led to the development of the local ecosystem, which includes co-working facilities, informal interaction centres (like coffee shops and cafés), testing facilities, etc. Bengaluru, too, has some areas that see a congregation of new ventures. In many cases, these are concentrated in one field or technology.

Other major cities in India—notably Chennai, Hyderabad, Kolkata and Ahmedabad—are also witnessing greater interest in entrepreneurship and innovative new start-ups. Long known for its entrepreneurial culture and business acumen,

Ahmedabad has been a laggard in the technological arena. This is surprising, given its history. The first spinning and weaving company was set up in the city, by Ranchhodlal Chhotalal, as early as 1859. He is reported to have asked the city to withdraw its support for a high school in 1886 and instead finance technical education. Apparently in keeping with this, from 1889 the city financed scholarships for technical students. Also, instead of just a few entrepreneurs introducing new industrial machinery, as elsewhere in India, in Ahmedabad the mercantile class as a whole supported the new techniques. Of course, this resulted in changes and affected the hand spinners and handloom weavers. However, many of them were recruited into the new manufacturing plants.

Ahmedabad is unique even within the diverse distinctiveness of India, with its very founding based on something special. As legend has it, at the beginning of the fifteenth century, an independent Sultanate ruled by the Muzaffarid dynasty was established in Gujarat. Sultan Ahmed Shah, while camping on the banks of the Sabarmati river, saw a hare chasing a dog. The sultan was intrigued by this and turned to his spiritual advisor for an explanation. The sage noted that a land where a timid hare chased a ferocious dog must be extraordinarily unique. Impressed by the strange incident, the sultan, who had been looking for a place to build his new capital, decided to found the capital there and named it Ahmedabad.

Despite its early adoption of technology in the mid-nineteenth century and the emphasis on technical education right from that time, Ahmedabad was bypassed, almost completely, by the IT industry. Only recently is it seeing some activity in this area. One could attribute this largely to two factors—the limited number of engineering institutes in the city (and in Gujarat as a whole), and the comparatively low level of diffusion of the English language. The latter demonstrates how state policy can have a serious impact. For a few decades, the state actually discouraged English in its schools. As a result, a whole generation grew up with but a small proportion of people fluent in English. This was a major impediment to the growth of any industry like IT, which focused on Western markets, especially when other locations in India had large pools of English speakers. It also inhibited the growth of technical education, which depends on textbooks in English (and therefore, familiarity with that language).

Many also attribute the absence of engineering education to the 'business' orientation of the educated Gujarati. This begins at an early age; apparently, even students in school dabble in stocks and shares, on the side. Personally, my earliest lesson in the 'what's in it for me' business culture of the typical Ahmedabadi came on my very first journey to Ahmedabad. I was travelling by an overnight train from Bombay, and an obviously experienced fellow traveller suggested that I go to sleep (the train arrived at Ahmedabad rather early in the morning), waking up only when the train halted at a station. He then explained to me, with great seriousness, what he called the 'traditional' way of knowing that one had reached Ahmedabad. 'At every halt, if you don't see the name of the station, just ask someone on the platform which city it is,' he said. 'If you wake up at each major stop, you will be told it is Surat, Baroda, etc. However, instead of responding with the name, if the person on the platform says, "Give me ten rupees to tell you", then it's Ahmedabad.' Certainly, an appropriate lesson—just before joining a business school—about the 'value of information'! And of course (even if unfairly exaggerated), about the business orientation of the typical Ahmedabad resident.

Another important factor holding back Ahmedabad is that in the 1980s and '90s, Gujarat was not very liberal in permitting private institutions of higher education. In contrast, the southern states and Maharashtra saw a huge boom in private institutes for engineering, resulting in a virtuous cycle between the growth of engineering education and of the IT industry. Also, there was a perception that the general standard of higher education in Gujarat was low. In large measure, this too was due to the neglect of the English language at the school level (the lack of high quality university-level textbooks in local languages meant that a good education required adequate mastery of English). Yet, strangely, Ahmedabad has long boasted of some of the best, even globally recognized, institutions of higher learning, like the Indian Institute of Management (IIM), the National Institute of Design and the School of Architecture (now CEPT University). It is also home to first-rate research institutions like ISRO's Space Applications Centre, Physical Research Laboratory, and Institute for Plasma Research.

There is no obvious answer to the conundrum of how a city with otherwise low educational standards is home to such world-class organizations. One factor may be the social ethos of the city, particularly its openness to outsiders. Though not a cosmopolitan city like Bengaluru (one indication is signboards and roadside advertisement hoardings; it probably has the smallest proportion of these in English, as compared to other big cities in India), it is regarded as a warm, friendly and safe city. Ironically, despite a history of horrifying, if only occasional, communal riots, in 'normal' times, it is probably the safest city in the country. Women feel secure on its streets even at night, and it is common to see girls and young women, bedecked in jewellery and their best clothes, out on the streets very late at night, especially at the time of the Navratri festival. Despite chaotic traffic and the inevitable scrapes and accidents, road rage is practically unknown (in sharp contrast with Delhi). Its normally peaceful and non-violent outlook is possibly the result of the influence of Jainism and Gandhism, combined with the Gujarati penchant to get on with business. Despite its parochial, as opposed to cosmopolitan, culture, outsiders who have come to Ahmedabad find that the practical charms of the city grow on them, and many settle down in the city even post-retirement.

Given its advantages, one can only conclude that these are overshadowed by its drawbacks. However, it seems likely that in the new wave of growth—especially in the technology and Internet-related areas—Ahmedabad will leverage its advantages to ensure it is no longer left behind. A social ethos that looks favourably upon entrepreneurship; the growth in engineering education; the innate business sense within the local community; a positive attitude towards risk and, therefore, innovation; the availability of venture capital; and the supportive policies of the government—together, these form a very conducive ecosystem, which can propel Ahmedabad forward to become one of the important hubs for innovation.

Another city with potential for innovation is Kolkata (earlier Calcutta). Much like Mumbai, it is a city born out of the British Raj. Earlier, it was but a village, and the capital of Bengal was Murshidabad, around 200 km north of Calcutta. In 1690, Job Charnok, an agent of the East India Company, chose the area for a British trade settlement. The site was carefully selected—it was protected by the Hooghly River on the west, a creek to the north, and by salt lakes about two and a half miles to the east. There were three large villages along the east bank of the Ganges, named Sutanuti, Gobindapur and Kalikata. These three villages were acquired by the British from the local landlords. In 1717, the Mughal emperor granted the East India Company freedom of trade in return for a yearly payment of Rs 3,000.[10]

In 1772, Calcutta became the capital of British India, and the first governor general, Warren Hastings, moved all important offices from Murshidabad to Calcutta.[11] Till 1911, Calcutta was the capital of India, when the British moved the capital city to Delhi.

From the 1850s, Calcutta witnessed rapid industrial growth. Trade too grew and, with it, the Calcutta port. Later, jute mills were supplemented by modern industries, and the region became the centre of the engineering industry. Coal and iron-ore mines were in its vicinity and the early integrated steel plants were in the east. India's first automobile factory was set up near Kolkata and it was a thriving industrial centre, with the added advantage (unlike Bengaluru, Delhi and Pune) of being a seaport.

Given the strong engineering-industry base and the fact that it was long famed as a centre for culture and education, one would assume that it would have been an ideal base for new technology industries. Like Ahmedabad, it missed the early stages of growth of the IT industry. In this case, the reason had more to do with perceptions about its social and political ambience rather than educational factors. Beginning from the 1960s, Kolkata became famous for labour strikes and dharnas

---

10. Kolkata, *Encyclopedia Britannica*, 2016.

11. Kolkata, *Encyclopedia Britannica*, 2016.

(practically a siege). A combination of strikes and government policies (like freight equalization, which nullified a great deal of the natural advantage of having mines nearby) led to a decline of industrial activity in and around Kolkata, and soon company after company began to close down. From the 1960s, the leftist movement in Bengal spawned a growing fringe, and extremism took root. Called Naxalism, after Naxalbari in north Bengal, where it began, this violent movement aimed at overthrowing the government. Drawing from communism and anarchism, the movement swept Bengal for many years and caused immense disruption. At its peak, it attracted to its fold armies of educated youngsters, and Kolkata—the intellectual centre of the movement—became a city of frequent bandhs (enforced shutdowns) and sporadic violence. Strong police action finally brought the situation under control, but the underlying disaffection, caused by inequity and exploitation, remains, and even today a large swathe of central and eastern India is dominated by 'left-wing extremism' or Naxalism. Though the mainstream Left opposed the Naxalites, militant trade unionism seeped strongly into Bengal. The constant disruptions of day-to-day life due to bandhs, along with severe power problems and a decaying infrastructure, made Kolkata an unattractive destination for any industry.

From the late 1990s, the situation began to change and Kolkata began to actively seek investment and attract new industries. A comfortable power situation and better infrastructure is now visible; the newer areas on the outskirts of the city are abuzz with activity and include many large IT companies. The historical baggage of the past—particularly of disruptive trade unions and general inefficiency—has not quite gone away in people's perceptions, but a new reality is slowly taking shape.

Kolkata has long regarded itself as the artistic, cultural and literary capital of the country. The cinematic achievements of its stalwarts, its famed writers and its fondness for intellectual debate make it a centre of creativity. As Gopal Krishna Gokhale, one of the top social and political leaders during the Indian Independence movement, said, 'What Bengal thinks today, India thinks tomorrow.'

It does, like Ahmedabad or Chennai, have a strong, but comparatively parochial (as opposed to cosmopolitan) culture, notwithstanding small modern/'Western' enclaves like Park Street. Yet, in the 1950s and into the '60s, Kolkata was very much the 'happening' city for young Westernized urbanites. Mumbai was way behind, and Delhi was not even in the running. However, as Naxalism and bandhs took over, the face of the city changed and its veneer of modernity quickly wore off.

Also, despite the one-time domination of local business by Marwaris from faraway Rajasthan, and a large labour force from Bihar, its demographic diversity is limited. Yet, the middle class has a passion for the artistic and literary—extending across cultures and countries—arguably unmatched elsewhere in India. Its unique 'addas' (local informal gatherings) are known to go beyond gossip and spur deep intellectual discussions.

With this ethos, and a background of so many decades of engineering industry, Kolkata is certainly well-positioned to be a leading centre of innovation. Once the capital of India and a city with a small but thriving cosmopolitan culture, Kolkata would certainly like to recapture its glory days.

Like Kolkata, Chennai takes culture very seriously. The crowds and enthusiasm at its music sabhas have to be seen to be believed. The rapt silence during performances is one indication of the seriousness and interest of the audience. Learning classical music or dance is almost compulsory for every middle-class youngster, especially for girls.

The city has a large film industry, and cinema artistes are adulated as nowhere else in India. Unlike the Bengali film industry, Tamil films are rarely focused on the artistic and are generally of the commercial variety. Why this medium of expression does not reflect the deep artistic interests of the middle-class Tamilian is difficult to say.

From being the major southern metropolis, the 1990s saw Chennai being overshadowed by Bengaluru, once considered a sleepy pensioners' paradise. The rapid growth of Bengaluru and its global visibility, thanks to its success as the capital of the glamorous IT industry, seemed to push Chennai off centre stage. Even emerging Hyderabad, with its dynamism and its can-do administration, was drawing away a lot of investment. However, the basic strengths of Chennai—its professional talent pool, its seaport location, attractive cost structure and hard-working labour force—ensured that it was very much in the running. In recent years, it has grown its manufacturing base and is now a major centre for the automobile industry, amongst others. It has attracted ever more IT companies and is known for the plentiful availability of engineering graduates, thanks to the rapid growth in engineering education in the state. The culture of hard work and diligence seems embedded in its young graduates, and this has made it an attractive centre for the BPO industry. As a result, it continues to be a major centre for the tech industry and *the* hub for manufacturing in the south.

Chennai's culture of meticulousness and sincerity is

exemplified by how delegates respond to conferences in different cities. In Delhi, a 9.00 a.m. start-time means that one can expect attendees to begin coming in by 9.15, and most will come in only by 9.30 a.m. (even that may be optimistic in winter!). In Mumbai, most will come by 9.15 and bemoan the traffic as the cause for delay. In Chennai, a majority will not only be in their seats by 8.45 a.m., but before 9.00 a.m. they will have their notebooks open and pens ready (or laptops booted up) to take notes, and will be awaiting the first speaker!

Times have changed, but Chennai once epitomized the ideal of simple living and high thinking (though the Kolkatan too would like to lay claim to that title). The 'Tam-Brahm' (Tamil Brahmin) was long famed for intellectual prowess and a severe, almost ascetic, lifestyle. Their ability to deal with numbers is legendary. I have often argued that the structure of the Tamil language (which, unlike other Indian languages, does not owe its roots to Sanskrit) hardwires the brains of Tamilians in a way that makes them adept with numbers. Similarly, Hungarian—the one European language not deriving from Greek and Latin—seems to hardwire its native speakers to be great numbers people. It is not surprising that a small country like Hungary produces a disproportionate number of top mathematicians. The uniqueness of the structures of both Hungarian and Tamil may possibly be the catalyst for producing mathematical geniuses.

The strong industrial and IT base, the abundance of engineering talent and the special aptitude for numbers provide Chennai with a unique combination of factors. These and other factors are likely to continue to make the city an attractive destination for investment in a wide range of industries, particularly those related to engineering and technology. However, from the viewpoint of innovation, the culture of conservatism and aversion to risk-taking amongst the traditional middle class are major constraints. The city is generally seen as a bastion that preserves and sustains classical culture, rather than one that experiments or drives change. Over time, many of the old values have, of course, changed and the new generation approaches life in a very different way. It is therefore possible that the impediments to innovation no longer exist. Yet, it seems unlikely that in the immediate future Chennai will be amongst the leaders in innovation.

An earlier chapter discussed Andhra Pradesh and the rapid rise of its capital, Hyderabad, as an important IT centre. In the days of the British Raj, the city was the capital of the eponymous state, considered the largest and the senior-most princely state. Its ruler, the nizam, was accorded a twenty-one-gun salute under the elaborate protocols of the Raj. Famed for his collection of jewellery, the nizam was certainly amongst the richest persons in the world. According to the Forbes 'All Time Wealthiest List' of 2008, the nizam (Mir Osman Ali Khan) is the fifth richest man in recorded history, with an estimated worth of US$ 210.8 billion (adjusted for exchange rates and US GDP growth).[12]

In the sixteenth and seventeenth centuries, Hyderabad was a centre of diamond trade. At that time, Golconda was the capital and famed for its fort. In 1589, the city of Hyderabad was built on the Musi River, about 8 km east of Golconda. In earlier times, the area was part of Ashoka's Mauryan Empire, and after his death (232 BC), the Satavahanas took over. Later, various Buddhist and Hindu kingdoms ruled the area. The

---

12. 'Making money the royal way!', *The Economic Times*, 23 April 2008.

Delhi Sultanate and the Bahamani Sultanate were in power in later centuries, followed by the Qutub Shahi dynasty (which built the city of Hyderabad).

The Asaf Jahi dynasty was founded in 1724, following a period of transition resulting from the erosion of the Mughal Empire's authority after Aurangzeb's death in 1707. Asaf Jah's successors ruled as nizams of Hyderabad, cleverly switching alliances between the French and the British (each then contending to spread its hold over the country) at different times. The city of Secunderabad (now effectively a part of Hyderabad) was first founded to station French troops. Later, British troops took over and they stationed a resident in Hyderabad.

When India won independence in 1947, the nizam declared his intention to remain independent. However, the law-and-order situation soon deteriorated, with people fighting to join the Indian Union. Finally, in September 1948, the Indian Army moved into Hyderabad and it was integrated into India. Later, in the 1956 reorganization of states, the erstwhile state of Hyderabad was divided between Bombay (later Maharashtra), Karnataka and the newly created state of Andhra Pradesh, of which it became the capital. Following the bifurcation of Andhra Pradesh in 2014, Hyderabad has become the capital of the newly created state of Telangana.

Hyderabad, as noted earlier, was vigorously promoted (especially from the late 1990s) as an investment destination. Apart from the IT industry, the state also promoted it as a centre for educational institutions. In addition to Osmania (a state university), it also has an IIT, the International Institute of Information Technology, the Indian School of Business and a large number of private educational institutions. It has, for many years, been home to a number of research and development institutions, particularly those related to defence.

The erstwhile Andhra Pradesh, especially Hyderabad, has emerged as a major centre for coaching classes that tutor students for the entrance examinations to professional courses. It has been eminently successful in this, and a substantial proportion of those who get admission to the premier institutions (through the very competitive entrance tests) are from this state.

As a result, the availability of talent is one of the assets of the state. In addition, Hyderabad is a reasonably attractive city for migrant talent. This, and the quick pace of infrastructure development, makes it a potential magnet for investment. Further, the positive policies and quick decision-making of the state government add to its attractiveness for both corporates and entrepreneurs.

These factors and the cultural diversity of Hyderabad (deriving from its history) should qualify it as a front-runner in innovation. Yet, it has so far not demonstrated the same kind of energy in the start-up space as Bengaluru, Mumbai or Delhi/NCR. As a result, the snowballing effect is not yet visible, nor has the innovation ecosystem developed as much as in the leading cities.

Another interesting southern city is Kochi. Despite its small size (compared to the cities discussed earlier), it is a particularly fascinating city, mainly because of its historical heritage and the diversity that has resulted. It is the biggest city and the commercial capital of Kerala. Like Bengal, this state too has long had a strong communist movement and powerful trade unions. It was perceived as one where endless strikes take place and as having a laid-back work culture. Never much industrialized, investors continued to stay away from the state.

The economic scenario began to change with the opening up of job opportunities in the Gulf area and the growing migration of Keralites to that region. Remittances from these migrant workers enriched their families and, as the numbers grew, had an impact on the state's economy. Interestingly, often considered easy-going and lazy at work within the state, the migrant Keralite established a reputation for hard work and sincerity abroad! This is not surprising, though, as, in earlier years when Keralites moved elsewhere in the country for employment, they were similarly in demand for their dedication and efficiency in the workplace.

With limited natural resources but immense natural beauty, Kerala wisely embarked on a mission to attract tourists. Its 'God's Own Country' campaign has been phenomenally successful and from being an unmarked location on tourist maps, it has become amongst the most popular destinations for both domestic and foreign tourists. Interestingly, it positioned itself as an attractive tourist spot for high-end visitors and not for the kind of backpack tourists who, at one time, flooded Goa. As a result, tourism became an important contributor to the state's economy. Between this and the remittances from the hundreds of thousands of migrant workers, the economy has picked up and the state is no longer a laggard on economic indicators.

Interestingly, despite long years of unimpressive economic growth, the state was, and continues to be, a leader in social development. On key health and education indices, its performance has been exceptional. The Kerala model of development, with its emphasis on social development, generated much debate and, in many circles, much praise. In recent years, though, it seems to have broken out of the controversy of social versus economic development by achieving growing success on the economic front as well.

Even a cursory look at figures tells the story of Kerala's high level of social development.[13] Its life expectancy at birth was 73.6 years in 2001–05 and 74.8 years in 2009–13, both being the highest amongst all states and substantially above the all-India figures of 64.3 years and 67.5 years for the two periods. Its infant mortality rate was just 12 (per thousand live births) in 2013. Its literacy rate has been the highest in the country from 1951, and was 94 per cent in 2011. Its social indicators can, in fact, be favourably compared with those of many developed nations. Of late, it has also focused on growth of industry and of its economy.

The achievements of Kerala in literacy, education and healthcare are certainly the result of enlightened policy initiatives and the priority given to these areas by the state. However, they are also due, in no small measure, to the social and cultural ethos of the state. This in turn has been shaped by the history of the state and the socio-political movements over the last century or more.

Centuries ago, its culture was influenced by the fact that it was a major trading post, resulting in interaction with people and cultures from across the world. Much of this was based on sea trade, for which Muziris was the main seaport, dating back to at least the first century BC. It was the node for trade with the Phoenicians, Egyptians, Greeks and the Roman Empire. Exports from Muziris were spices, semi-precious stones, pearls, diamonds, sapphires and ivory. Imports included gold coins,

---

13. *Economic Survey 2014–15* (Volume II), Ministry of Finance, Government of India, February 2015.

linen and clothing, wine, tin, copper, etc. Later, from the fifth century AD, trade with Rome was replaced by that with others, particularly the Chinese and Arabs.

The exact location of Muziris is still a matter of speculation, but it is generally thought to have been near present-day Kodungallur (formerly Cranganore), which is about 30 km north of Kochi. Muziris disappeared from early maps, without a trace, presumably because of a cataclysmic cyclone and floods in 1341. This is said to have altered the geography of the region, leading to the death of the port.[14]

Thomas, the disciple of Jesus, is said to have arrived in India through Muziris in AD 52, and converted some prominent local families. Thus, Christianity arrived in Kerala (and India) before it did in much of the West—a fact that is not widely known. Some records also indicate that Pantaenus, a leading theologian from Alexandria, visited Kerala at the invitation of the local Christian community around AD 190.

Thus, Christianity has been in Kerala for almost two millennia. If that comes as a surprise, not many may know that Jews in Kerala pre-date even the arrival of Christianity: they are said to have come to Kerala in 587 BC, when they fled the occupation of Jerusalem by Nebuchadnezzar. Today, the synagogue in Kochi is mainly a tourist destination and is rarely used for religious purposes. Rituals like a public prayer require a Minyan, a quorum of ten Jewish adults (thirteen years or above). With a large number of Jews having migrated to Israel, such a quorum is now difficult, particularly as Orthodox Judaism does not include adult females in the quorum number. The 2001 census recorded only fifty-one Jews in the state, and

14. Muziris Heritage Project website.

the number is reported to have since declined. Yet, doubtless, Jews too have contributed to the multiculturalism of Kerala. The broader religious diversity has continued and amongst the states, Kerala is unique in its large proportion of three different religions, with a population that comprises 54.7 per cent Hindus, 26.6 per cent Muslims and 18.4 per cent Christians.[15]

Reflective of the diversity of the state is the history of its premier city, Kochi (formerly Cochin). The seat of the eponymous princely state, Kochi traces its history back many centuries. Trade with Greeks, Romans, Jews, Arabs and Chinese made it an important commercial centre from the fifteenth century, prior to which its history is not well-documented, except that Cochin State is said to have come into existence in 1102, after the breaking up of the Kulasekhara empire. Subsequently, the raja of Kochi remained the titular head through the years of foreign control: first by the Portuguese (1503–1663), then the Dutch (1663–1795) and finally the British (from 1814 to 1947). In between, Hyder Ali of Mysore conquered parts of the kingdom in 1773.

Kochi arrived on the world trading map after the destruction of the port of Muziris in 1341. The Portuguese set up a factory in Kochi, and got the raja to build a protective fort around it (Fort Manuel, now Fort Kochi). This became their base and Kochi was the capital of Portuguese India till 1510. After various encounters, the Dutch defeated the Portuguese, beginning their 132-year rule over Kochi in 1663. Under them, Kochi prospered, mainly by trading pepper, cardamom and other spices, coir, coconut and copra. In 1814, under the

---

15. Census 2011 figures, as quoted in *The Times of India* (Delhi), 26 August 2015.

Anglo–Dutch Treaty, the islands of Kochi and its territory were ceded to the United Kingdom.

As trade through the port increased, the British felt the need to develop it further. Lord Willingdon, the then governor of Madras, brought harbour engineer Robert Bristow to Kochi in 1920 to determine how the port could be expanded. To create a new port at Kochi, he dredged extensively (reports indicate that a dredger was used for twenty hours a day for two years), resulting in the port (opened in 1928)[16] being able to receive vessels that needed up to 30 feet draught.[17]

The dredged sand was used to create a new island with an area of around 3.2 sq. km. It was named Willingdon Island, after Lord Willingdon who had then become viceroy of India. It is now home to the office of the Port of Kochi, the Customs office, as well as a naval base of the Indian Navy. The synergistic creation of the port and Willingdon Island is a good example of innovative thinking.

Although they witnessed the arrival of a large number of foreign traders and conquerors in earlier centuries, Kochi and Kerala no longer see that kind of influx. In recent decades, Kerala's slow economic growth has meant limited job opportunities; therefore, migration of outsiders (even from within India) into the state has been limited. However, the big tourist influx—from elsewhere in India and around the world—and the large number of returning and visiting emigrants have continued to introduce the state to other cultures, languages

---

16. '75 years ago, Cochin Port happened', *Metro Plus Kochi*, *The Hindu*, 26 May 2003.

17. 'Cruising With Five-Star Hospitality', PIB newsletter, 20 November 2002.

and societies, virtually replicating a high-diversity society. Further, a 2016 report estimates an increasing influx from other parts of India, with the annual number of migrants arriving for work in Kerala being around 235,000.[18] Of these, 75 per cent are from the eastern states of West Bengal, Bihar, Assam, Uttar Pradesh and Odisha. A direct train, the Vivek Express, from Dibrugarh in Assam to Kanyakumari, the southern tip of India (in Tamil Nadu, but just adjoining Kerala), has greatly facilitated this movement. Incidentally, the 4,233 km route takes eighty-three hours and is the longest rail journey in India.

On the basis of the above discussion, one can assess the cities identified as centres of innovation. The key features of each city are summarized below.

Bengaluru

The city scores high on talent availability, its ability to attract and retain outside talent (including expatriates), and the existing base of start-up ventures. The support system and institutional presence is good, as is the availability of risk capital. The physical infrastructure is greatly strained and clearly unable to cope with the rapid growth of the city, but connectivity and location are positives. The cosmopolitanism, social infrastructure and climate make for a generally welcoming ambience and contribute positively to the quality of life. State policies have ranged from good to extremely positive, and the overall social ethos (despite occasional bouts of regionalism and prudishness) is helpful.

---

18. 'The Big Shift: New migrant, new track', *The Indian Express*, 15 August 2016.

Mumbai

On availability of a large talent pool, Mumbai scores well. Those living there seem to appreciate the positive features of the city and its ability to retain talent is good. Attracting new talent, though, is not so easy, since the city is known to have a high cost of living and time-consuming commuting. Yet, the prospects it offers are an attraction, as are its cosmopolitanism, efficiency and strong social infrastructure. The diversity of the city and the availability of risk capital are big positives for innovation, as is the support infrastructure needed for a start-up. The presence of creative ventures in different fields (notably film, fashion and art) is another positive.

Pune

A strong engineering base, good talent availability and proximity to Mumbai are the main advantages of Pune. As a centre of culture, it has strong inherent creative energy. However, on diversity and availability of risk capital it does not, at present, score very highly. The institutional infrastructure is good, especially in the area of technology. The number of start-ups is growing, but is not comparable to Bengaluru.

Delhi/NCR

A late starter, especially with regard to technology, Delhi/ NCR has moved forward quickly. Its ability to attract outside talent has improved, and the snowballing effect has now made it a magnet for talent not only from its vast hinterland, but from all over India. The character of the city and its surrounding towns has undergone a transformation and it is abuzz with entrepreneurial activity. Through it yet lacks the extensive nightlife of Mumbai, it has a well-developed social infrastructure. Personal safety, especially for women, is perceived as a problem and is the one deterrent for outside talent.

Chennai

Through a major manufacturing and IT centre, Chennai does not yet have a large start-up base. The very large pool of talent, especially in the field of engineering/technology should, at some point, propel it towards innovation, and proactive policies may well help to catalyse this.

Other Cities

Hyderabad, Ahmedabad and Kolkata have some positive features, but are yet to evolve the overall ecosystem that would make them major centres for innovation. Jaipur and Chandigarh are yet to develop any substantial base, though both have been active in trying to attract the tech industry. Kochi, which was discussed earlier, is yet to reach the scale that is necessary, though many initiatives there, aided by a proactive state policy, are indicative of future potential.

INNOVATIVENESS STATUS AND POTENTIAL

At present, there is little doubt that Bengaluru is the city that stands above the rest in terms of entrepreneurship based on innovation. Mumbai, Pune and Delhi/NCR follow, in that order, though the contest between Pune and Delhi/NCR is close and the latter could as well be adjudged ahead of Pune. The next tier of cities is considerably lower, with innovative entrepreneurial ventures being in far smaller numbers. Differentiation amongst them is yet to emerge. A reasonable assessment would, it seems, place Chennai first, then Kochi ahead of Hyderabad, followed by Ahmedabad and Kolkata. Other cities are yet to truly enter the game.

Over the next few years, the need to innovate, driven by competition in the marketplace, is going to increase. At the same time, the possibilities will also increase considerably, driven by technological advances. Facilitating innovation will be policies, changing and supportive societal values, and easy availability of risk capital. Overall, the ecosystem for innovation will develop and be far more conducive than it is today.

Within this broad framework, there could be differences amongst cities on the various factors that drive innovation. Given that risk capital travels easily, this is unlikely to be a major differentiator. However, the guidance and mentorship that goes with it tends to be geography-specific, with the location of mentors being a positive or constraining factor. Similarly, while the formal policies of various states may tend to converge—as they learn from each other—and provide incentives that are similar, actual implementation will vary, creating differentials amongst states.

The availability of suitable high-quality talent in sufficient number will continue to be a necessary condition for innovation. This, as noted earlier, includes the ability of the city to attract outside talent and to retain talent, and depends on various factors, of which two are particularly important. The first is physical infrastructure of the city, a factor that is also otherwise an important facilitator of innovation. The second is the social or 'soft' infrastructure. Together, these two also define the quality of life in a city.

A key influencer of innovation will continue to be the culture and values of the city—what has been called its social ethos. The general belief is that this, like all social evolution, can change only slowly, over decades, if not generations. Yet, there are instances where the change has been rapid, influenced by a sudden surge of immigrants or by other events (I mentioned Gurgaon as an example of such change). Sometimes, the old culture may continue, but will coexist with a new and very different one (Pune being an example of this). Therefore, in looking at the future, one cannot assume an unchanging social ethos.

Given the uncertainty in many of the key factors, trying to rank cities in terms of their innovativeness in future is a hazardous exercise. Yet, given its lead, its present momentum and the likely scenario, it is a fairly safe prediction to say that Bengaluru will continue to be number one till 2020.

Within the same time horizon (looking at a longer time frame brings in too many uncertainties), we do not see a substantial change in the broad grouping of other cities that were assessed earlier. At the second tier, after Bengaluru, would be Mumbai, Pune and Delhi/NCR. In all probability, Delhi/NCR will overtake Pune. Driving this change will be the easier availability of risk capital, a more positive societal attitude towards entrepreneurship and risk, and better physical infrastructure.

Mumbai will stay ahead of Delhi because of the greater number of start-ups and the resulting facilitative ecosystem. The general efficiency of work, the energy and the greater creative ambience are other factors that will help it stay ahead.

Other cities have not been discussed here. In our assessment, the large cities in the north—barring Delhi and its environs— are unlikely to be anywhere near as innovative as the cities mentioned earlier. Thus, we do not see Patna, Varanasi, Allahabad or Lucknow as big centres of innovation in the near future; nor even Kanpur, despite its past industrial history. Chandigarh has tried, for some years now, to become an IT hub, but its success in this has been limited. In Madhya Pradesh, Bhopal and Indore (and Gwalior too) have aspirations, but they are likely to remain just aspirations, at least till 2020, even though Indore is shaping up well, if slowly. The same prognosis may be made of Nagpur and Nashik in Maharashtra. Vadodara and Surat (a large and booming city) in Gujarat also do not show any indications of having special potential for innovation, despite Vadodara's artistic and cultural legacy or Surat's tremendous entrepreneurial acumen and growth (the city, famed for its diamond industry, is now the tenth largest in the country).

The newly created state of Andhra Pradesh is promoting Visakhapatnam (popularly known as Vizag) as a centre for the IT industry, and has entered into a Memorandum of Understanding with the founders of Kochi's Startup Village to create a similar facility. Despite the proven dynamism of the state's Chief Minister Chandrababu Naidu, it is unlikely that Vizag will become a major centre for innovation till 2020.

Another state that has been proactively promoting entrepreneurship and innovation is Kerala. Apart from the Startup Village initiative, the government is encouraging student entrepreneurship through start-up boot camps in engineering colleges. It has also introduced a gap year for student entrepreneurs as well as grace marks and flexi-attendance. Its Technology Startup Policy 2014 aims to incubate 10,000 technology product start-ups by 2020 and become one of the top five start-up ecosystems in the world.

Thus, in our assessment, it is unlikely that we will see any of the other cities being in the same league as those in even the last of the three categories of innovative Indian cities. This means that even in 2020, the three categories together will not expand to a 'Top Ten'.

Within the nine, will there be any dark horses that suddenly nose ahead? Might a city like Kochi, with its innovative Startup Village, or Ahmedabad with its entrepreneurial drive, become a contender to enter the top league of innovative cities? What about the other big metros, Kolkata, Chennai and Hyderabad—might one of them not take a big leap forward? The uncertainties of the future preclude a definitive answer to this question. The prediction is, therefore, tentative. Based on an assessment of how innovation-influencing factors might evolve, I would assert that neither Hyderabad nor Chennai will join the category of second-level innovation centres (Mumbai, Delhi/NCR, Pune). Both are well-placed with regard to talent, but the other elements influencing innovation are not yet at the level or scale that will propel them to the top league. Comparatively, I feel that, looking ahead, Chennai has more potential than Hyderabad. Kochi, despite many positives, will not have the scale to become a major centre. Its excellent boot-strapping efforts are succeeding, but are limited to the field of technology. Kolkata remains bedevilled with too many systemic inefficiencies and the overall ecosystem will take time to evolve to a greater maturity. While its creativity quotient is high, its business or entrepreneurship one is low. Ahmedabad has an almost diametrically opposite situation with regard to creativity and entrepreneurship. Its ecosystem is evolving fairly quickly and risk capital availability is positive. However, talent continues to be its Achilles heel. Changing this is, unfortunately for that city, not something that can be achieved in the short term. Over a more extended time frame, by 2025, Ahmedabad could well become a major hub for innovation.

In keeping with this brief assessment, we would place Chennai at the top of this third tier of cities. In due course, it may well join the category of Mumbai, Delhi/NCR and Pune. After Chennai, we would place Ahmedabad and Kolkata—very different cities, with many polar opposites between them. A combination of their complementarities would place them in the top league! Finally, despite their achievements—the phenomenal rise of Hyderabad, the uniqueness and success of Kochi's Startup Village—these two cities are left at the end of the table.

A summary of present and future ranking is as follows:

Present + (2015)	Future (2020)
1. Bengaluru	1. Bengaluru
2. (a) Mumbai	2. (a) Mumbai
(b) Pune	(b) Delhi/NCR
(c) Delhi/NCR	(c) Pune
3. (a) Chennai	3. (a) Chennai
(b) Kochi	(b) Ahmedabad
(c) Hyderabad	(c) Kolkata
(d) Ahmedabad	(d) Hyderabad
(e) Kolkata	(e) Kochi

Thus, I foresee no dramatic change, with no category-leaps by any city. In the second category, Delhi/NCR overtakes Pune, but even in terms of present status, I had indicated the two as being very close. The only churn is in category three, mainly because the starting base in all five cities is low. Chennai will retain the top spot, but is unlikely to do enough to move to category two. Kochi drops considerably, mainly because it is unlikely to have scale and also because both present and future innovation is unidimensional, being concentrated almost solely in the tech space. Ahmedabad, Kolkata and Hyderabad all have a much broader industrial/commercial base and are likely to be able to quickly ramp up the facilitating ecosystem for innovation.

This chapter and the earlier one have identified and discussed the factors that facilitate (or retard) innovation at both the state and city level. While it has discussed how some states measure up against some of these factors, I have not attempted a ranking of the states, since many of the major factors that determine innovativeness are city-specific. Accordingly, I have looked at nine major cities and assessed how each of these measures up against the innovativeness factors that have been identified. It is important to note that the assessment is based on a specific interpretation of innovation: namely, the beneficial (commercial or non-commercial) use of a new or different idea. It is, therefore, linked to entrepreneurship. Based on this and my assessment, the cities have been ranked on innovativeness.

The following chapter looks at how organizations, both commercial and non-commercial, might increase their innovativeness. It also discusses how policies and actions—at the state, central or city level—can help to make a city more innovative.

7

FOSTERING INNOVATION IN ORGANIZATIONS

Innovativeness of organizations as determined by policies, structures, processes and organizational culture. Do size and sector matter? Promising innovations that failed. Encouraging risk-taking: ideas fail, people don't; water-cooler moments. Prizes, awards and grand challenges as stimuli for innovation.

The previous chapter assessed, and then ranked, some cities with regard to their present and future innovativeness. If a similar assessment were to be done for organizations, what criteria would one look at? What are the policies, structures, processes and organizational culture that differentiate innovative organizations from others? Are these similar across sectors, countries, and categories (government, not-for-profit, commercial)? What follows discusses these questions and related issues.

Defining an innovative organization is not as easy as it may seem. A good definition requires objective and preferably quantifiable indices or metrics, but identifying these is not a simple task. Some corporates use a single index—percentage of revenue derived from products or services introduced in the last three years. This has the advantage of being objective, as also of being easily quantifiable and comprehensible to all. One pitfall is the time period: many innovations take a long time to translate into sale and revenue and so a three-year horizon may be inadequate. On the other hand, a longer time period could result in one—possibly stray or accidental—blockbuster product, indicating strong innovativeness for many years, even if no other new product is developed. To avoid such misleading distortion, one could add a second factor: the number of new products or services commercialized in the last three (or five) years. This two-factor index is certainly a better indicator of innovativeness.

One difficulty in this methodology is in identifying 'new' products or services. In any ongoing venture, there are always continuous improvements being made to the products and services it offers. Therefore, it is not always easy to differentiate what is merely a tweaking of an existing product or service from one that marks a radical departure. Thus, while the final index is two objective numbers, they are based on a subjective judgement of what is 'new'.

The second and more fundamental problem lies in the very definition of innovation. The above approach takes a narrow view of innovation, limiting it to new products or services. It ignores innovations in the process, or fundamental changes in the business model. These, as discussed in earlier chapters, can be game changers that are potentially at least as important as a new product or service.

Clearly, measuring the innovativeness of an organization is not as simple or straightforward as it may seem. It needs a wider range of indices and at least a few of them will be subjective. At the extremes, it may be easy to categorize organizations as innovative or non-innovative, but in the vast area between these opposites, it is far more difficult. Thus, there is a general consensus, based on its phenomenal track record, that Google is innovative, and that Kodak died because it did not keep up with innovations, but the overwhelming majority of organizations lie in between, and judging their innovativeness is not a simple task.

One view is that inputs which are more amenable to measurement can provide a reasonable index of innovation (many country assessments follow this approach). These input indicators include expenditure on research and development, number of scientists/technologists and the proportion of space dedicated to research and development/innovation. This, of course, assumes a clear demarcation of the research and development/innovation activity, so that one can identify the budget, people and space dedicated for this. In many cases, such a neat separation does not exist. In fact, there is a strong argument that a lot of innovation emanates from those directly involved in the business, and not from an independent research and development facility. In any case, it is questionable as to how much of the input translates into actual innovation. A high level of inputs does not necessarily result in many innovations. On the other hand, a highly innovative organization may well produce many innovations even with low inputs.

The efficiency with which inputs are converted into innovation is a good way of identifying innovation-efficient organizations. The extent of input (as a percentage of revenue or profit or, in the case of a country, of GDP) is an indicator of the leadership's commitment to innovation, with input being the budget, people and space dedicated to research and development or innovation. Despite some difficulties (people who work part-time on innovation, or shared space and facilities), this is amenable to quantification. On the other hand, output in terms of innovation is not easily quantifiable. There are, in some cases, simple indices like number of patents filed or number of research papers published. However, apart from the inadequacies of these in capturing the full extent of innovation or its quality and significance, these do not cover innovations in business models or non-patentable process changes. Also, a patent does not indicate the extent of impact: a patent for a razor has significance that is vastly different from one for a life-saving drug.

Despite these difficulties, if one is able (even judgmentally) to assess the innovation output of organizations relative to the inputs for them, it would be possible to map out those that are innovation-efficient. One could then rank them on this basis, somewhat along the lines of what the Global Innovation Index does for countries. Such an effort is beyond the scope of this work, and is not attempted here. It is a task that academic institutions might like to take forward. However, even without an explicit ranking, one can identify a few innovative organizations and seek out the factors that make them more innovative. This is what is explored here.

The first and most important, as also the most obvious factor, is the organization's viewpoint: does it encourage innovation? On the face of it, there would be few organizations that do not claim to do so; but reality belies such assertions. Many organizations do not want disruptive change (which is, inevitably, what innovation causes) and often for good reason. An assembly-line factory depends on a fixed process and any change would require high investments in time and money in order to be validated for quality, reliability, repeatability, cost and time. It would, therefore, be resistant to suggestions for change, leave alone the radical change that innovation results in. Such an organization's culture is, quite naturally, focused on following rigid procedures, which is what ensures quality and repeatability, and deviations are not welcome. This is the mindset of most organizations in traditional industry. It is also seen in other long-standing institutional structures. Government is a good (or bad!) example: bureaucracy depends on fixed and known procedures, time-tested and ingrained over decades. Being a large system, the risk of a perturbation caused by something new is high. Change is, therefore, a slow and incremental process, with each step being small and carefully tested before it is accepted and integrated into the system. Obviously, the (often unstated) policy is passive discouragement of innovation. In general, large organizations with much at stake work on the basis of a risk-reward matrix, which is heavily skewed towards risk minimization.

In such organizations, continuity, stability and predictability are virtues. As a result, innovation is uncommon. Recognizing this, many of them try to separate and firewall smaller units, dedicated to innovation or research and development. Newer organizations—start-up ventures—do not have to worry about these factors, given their lack of legacy and history. This makes them more likely to be innovative. In a sense, the basic 'DNA' of the two are different. An organization that has grown rapidly is sometimes able to retain its innovation gene. One might argue that this is the case with Google. Even so, it too has set up a separate entity (Google X) to innovate and create new products.

In August 2015, Google went a step further. It created a holding company, called Alphabet, thus separating new, start-up and innovative activities from its bread-and-butter operations, which continue to be called Google. This organizational ring-fencing of innovative initiatives from routine business is clearly aimed at fostering entrepreneurial and innovation activities.

Another challenge for large organizations—even comparatively younger ones—is the result of their size. A big organization necessarily requires a degree of uniformity in procedures and systems. It needs reporting structures, hierarchy and checks and balances. Sooner rather than later, these become rigid (especially since predictability becomes important) with little leeway for flexibility or judgement. This is the essence of the bureaucratic system, with its focus on curbing arbitrariness, on being fair and equitable and on efficiency. Undoubtedly, this has many advantages in an organization engaged in routine and well-defined tasks.

It is, however, a major constraint for innovation. Large organizations have tried to get around this in different ways. One approach, as mentioned, is to create a separate unit dedicated to innovation. Another is to decentralize, so that virtual small organizations are created. Some have systems to encourage entrepreneurship within the organization ('intrapreneurship'), while others fund such internal entrepreneurs and give them a few years to create a start-up, which is then folded back into the mother organization. A few just acquire promising start-ups.

Unstated—and often unintentional—signals are a deterrent to innovation. A perception that the organization greatly values stability and continuity may well discourage innovation. Countering this requires an explicit policy that promotes new ideas. Many organizations do this through internal contests and recognition or rewards for the best innovations. This does promote innovation, especially if a message is sent that the top management is strongly behind such a contest and values innovation.

A few organizations go further. Recognizing that fear of failure is a major barrier to innovation, they not only recognize the most successful innovations, but also reward selected failures. Typically, these are ideas that were potentially game-changing but, for one reason or another, did not quite work out. Such acknowledgement of the risk of new ideas, that not all of them will culminate in success, conveys well the message of innovation. It encourages people to come up with truly out-of-the-box and radical ideas, and not limit themselves to safe, incremental ones that have a high probability of success. Strange as it may seem, rewarding failure is a good way to encourage innovation!

Such an approach—of rewarding even failed ideas—is particularly important in India, given the conventional mindset of looking down on failures. Given this, in organizations and in society, few are willing to undertake a venture that may well fail, and even limited success is seen as better than the chance of a failure. In a status-conscious society, peer recognition is very important and no one wants the lowering in esteem that comes from failure. Organizational reward for a 'grand failure' is therefore an attempt to change attitudes and ensure peer respect for a person who had a great idea, even though it might have failed.

The Tata Group has addressed and tried to overcome many of these issues. In the context discussed earlier, size is certainly one of its handicaps[1]. A conglomerate, it posted revenues of

---

1. What follows is substantially from 'Power of One' by Priyanka Sangani in *The Economic Times Corporate Dossier*, 7–13 August 2015.

$108 billion in 2014–15, and had a market capitalization of $134 billion. It has focused strongly on innovation and, in 2015, the top sixty innovations across the Tata Group were expected to deliver an estimated financial benefit of $1.1 billion annually. The group has set up a Tata Group Innovation Forum. Amongst its initiatives is 'Tata InnoVista', the in-house innovation awards, initiated in 2006. In 2015, there were 2,700 entries from seventy-five group companies from across eighteen countries. Collaboration across companies in the group is encouraged.

In the context of rewarding failure, the Tata Group has, as part of its innovation awards, a 'Dare to Try' category for audacious innovations that did not work. Entries for this grew from twelve in 2007 to 176 in 2014.

Venture capitalists in mature ecosystems (as in Silicon Valley) recognize this in a different way, clearly separating the idea and the person. A failed idea is unlikely to get support, but the person who conceived it is seen as having gained experience. Thus, a serial entrepreneur—even one with failed ventures—is more likely to get funding than one who has limited success to boast about. The former is not only seen as having gained in experience—at least about what not to do—but is clearly a person with many ideas, an innovator. Support and funding is as much—or more—for the person, as for the idea. The deeper message in this is important for innovation (and possibly, even beyond): ideas fail, people do not.

It is important to note that innovations, howsoever promising, do not always result in success. There is many an example of interesting innovations, with apparently great potential, not taking off. In 1999, a small group of enterprising experts from Indian Institute of Science created a handheld device intended as a low-cost computer, especially for rural India. The Amida Simputer, as it was called, incorporated an accelerometer—the first ever in a handheld device (a feature that became common after Apple's iPhone). In the early days of the Simputer, back in 2001, *The New York Times* noted: 'the most significant innovation in computer technology in 2001 was not Apple's gleaming titanium PowerBook G4 or Microsoft's Windows XP. It was the Simputer, a net-linked, radically simple portable computer, intended to bring the computer revolution to the Third World.'[2] Despite this ringing endorsement, the Simputer got no orders from the governments (state or central) in India and though it did make some sales in Africa, it never took off.

Another innovation, much hyped, partly due to its pedigree, is Google Glass. Announced with much fanfare in 2012, the sale of its test unit was reportedly stopped in early 2015. The spectacles-like device helps the user make and answer calls, send messages, take pictures and videos, get directions and video-chat. Google says it plans to reboot the wearable computer, which has apparently been moved out of Google X (the special innovation organization) and into a mainstream products division, before relaunching it.[3]

One can only speculate as to the reasons for the market's rejection of innovations that seem to have so much potential. In the case of Simputer, it may have been before its time. Also, as some have speculated, users may have found it too expensive compared to a cell phone for only a few additional features, and not versatile or capable enough to replace a laptop or

---

2. 'Simputer', *The New York Times*, 9 December 2001.

3. 'Google halts sales of Google Glass', *The Telegraph*, 15 January 2015.

desktop computer. Google Glass, many feel, was exorbitantly priced (US$1,500) and had limited applications. In addition, there were concerns about privacy.

Tata Nano was mentioned earlier, and certainly it was innovation—in fact, many innovations—that made possible this sub-one-lakh-rupee (Rs 100,000) car. While it got much worldwide publicity and acclaim, it did not do well in the marketplace. Some technical problems resulted in adverse publicity, but the biggest cause of failure was probably the positioning and marketing. The tag of 'cheap car' overlooked why many in India buy a car: not just for convenience, but also to enhance their social status. A cheap car was hardly the way to do this. Maybe positioning it as the best second car or the most convenient city car (especially given the traffic and parking problems in India) or a high-tech car may have worked better. As it stands, Tata had to drastically cut production, as demand fell precipitously. It is now being re-launched with many additional features and at a higher price point.

In what organizational milieu do new ideas best flower? In an earlier chapter, we discussed the role of freedom and democracy in promoting innovation. The same applies to organizations too. A centralized, command-and-control type of structure is not conducive to innovation. In such organizations, instructions, wisdom and ideas are seen as flowing from the top; those below are expected to execute these efficiently. In such a set-up, one can hardly expect ideas and innovations to flower.

On the other hand, organizations that are based more on shared goals, of motivation that comes from common objectives, tend to provide more autonomy to people. Authority and power are typically decentralized, with decision-making responsibility delegated as far down the line as possible. With such a structure, the organization is necessarily less hierarchical. Such organizations tend to provide a far better ambience for innovation. Organizational structure is therefore an important element that determines the innovativeness of an organization.

Processes and systems within organizations play an important role in determining employee behaviour. As noted earlier, with growth and time, there is a definite tendency for procedures to become standardized, bureaucratic and rigid. The expected organizational behaviour is oriented towards compliance and conformance. The procedural rigidities transform into behavioural rigidities. In such a situation, innovation will be minimal. Since there is little way out of standard procedures for any large organization, one way of tackling this problem is to minimize their importance. Innovative organizations focus on people and ideas, and not on procedures. A contrary example is best provided by the typical government organization. They are designed to be impersonal—with fixed positions, into which people come and go—and process or procedure-centric. This is good for routine administrative tasks, but is hardly likely to promote innovation.

An organization's working culture is determined by its policies, systems, structure and processes, amongst other things. However, it is also defined by a number of unstated, often intangible factors. What happens at the water cooler or during the lunch break also contributes to organizational culture. The seemingly trivial exchanges among colleagues, opportunities for interaction, walk-and-talk versus email— these are additional determinants of organizational culture and, in turn, influence the innovativeness of an organization. Cross-functional and trans-disciplinary interaction provides a vital stimulus for ideas, and is an important driver of innovation. While some organizations have formal mechanisms for this, what works best is a general environment that provides opportunities for such interaction. In this, the physical design of the organization is as important as the conceptual structure. Who sits where, the extent of isolation of different departments and of the leadership, the placement of pantries/cafeteria—factors like these can encourage or inhibit interaction. A boss who has to walk through the work area for every visit to the toilet, coffee machine or water cooler will inevitably have greater informal interaction with his/her team than one who does not really need to leave his/her cabin. Daily informal interaction of this type promotes the degree of comfort that makes the exchange of ideas—even far-out ones—easy and natural. Such organizations tend to be more innovative, since conversations catalyse and transform ideas, helping shape them into practical and viable products or services.

The world of venture capitalists and start-ups has spawned the new phrase, 'elevator pitch'. This is the quick pitch (for funding) made by an entrepreneur to a venture capitalist who s/he has as a captive audience for a few brief minutes while in the elevator. Similarly, an appropriately designed office may present a budding innovator an opportunity to make a 'water-cooler pitch' while the boss is at the cooler.

A few decades ago, in ISRO's early days (and before the concept of elevator pitches became popular), I experienced an interesting variant of the elevator pitch. I—and others—sometimes used the opportunity provided by a shared toilet to pitch new ideas to Dr Vikram Sarabhai (then chairman of ISRO) through a 'toilet pitch' or 'toilet tryst'!

Conversations and collaboration, especially across levels of hierarchy, are more frequent, meaningful and robust when the organization is comparatively democratic. Freedom of speech, acceptance of criticism (even by the leader), and respect for minority views are the hallmarks that characterize a democracy. Similar features within an organization are important drivers of innovation.

Conversations and collaboration, when carried out in a trans-disciplinary context, can be major stimulants of innovation. A solution or an analogy from one field for a problem in another is often an important source of innovation. Teams and organizations that work at the intersection of disciplines are more likely to come up with new ideas as the cross-pollination between varied expertise takes place. This is one reason why universities, with their varied fields of study, are the source of so much discovery and innovation.

It is, though, not enough to have varied disciplines within the organizational ambit; there is a need for them to interact as well. It is in this that the water cooler, cafeteria and common spaces play an important role. In many ways, the architect and interior designer play as important a role in promoting innovation as the CEO!

Organizations may not have too many different disciplines or experts within them. One way of simulating this is through diversity. This has, to some extent, become as much of a buzzword as innovation itself. However, being a cliché does not undermine the basic importance of either of them. Diversity goes beyond gender, nationality, expertise, age and other such obvious factors. One of its most critical facets is diversity of thought or approach, and this is key to innovation. Organizations that want to be innovative need to keep in mind the need for such diversity.

Some organizations follow a target-oriented or mission approach to innovation. Thus, one may define an end goal and then set a team (or teams) the task of achieving it. A new or difficult goal is likely to need innovation. An example of this is President John F. Kennedy's reaction to the Soviet Union's launch of the Sputnik I on 4 October 1957, a development that shocked the US, which had fancied itself as the world leader in technology. He announced that the US would put a man on the moon by 1970. This mission, with its specifically defined target (and a deadline) could only be met through numerous innovations in technology and systems.

Similarly, Tata took up the challenge of creating a sub-one-lakh rupee car. This goal necessitated innovations in technology, production, materials and management.

The idea of stimulating innovation by setting a goal has been adopted by various organizations through prizes for 'grand challenges'. Amongst the well-known ones is the X Prize. This is a non-profit organization that designs and manages public competitions intended to encourage technological development that could benefit mankind. The Board of Trustees includes Elon Musk, James Cameron, Larry Page, Arianna Huffington and Ratan Tata, among others.[4]

The X Prize mission is to bring about 'radical breakthroughs for the benefit of humanity' through incentivized competition. It fosters high-profile competitions that motivate individuals, companies and organizations across all disciplines to develop innovative ideas and technologies that help solve the grand challenges that restrict humanity's progress.

The first X Prize—the Ansari X Prize—was inspired by

4. Official X Prize website.

the Orteig Prize, a $25,000 prize offered in 1919 by French hotelier Raymond Orteig for the first non-stop flight between New York City and Paris. In 1927, Charles Lindbergh won the prize in a modified single-engine Ryan aircraft called the Spirit of St. Louis. In total, nine teams spent $400,000 in pursuit of the Orteig Prize.[5]

The Orteig Prize, though, was not the first of its kind. Over two centuries before it, in 1714, Britain's Parliament created a Longitude Prize of £20,000 as reward for finding a way for ships to determine their location within half a degree of longitude.[6]

In 1996, entrepreneur Peter Diamandis, founder of the X Prize foundation, offered a $10-million prize to the first privately financed team that could build and fly a three-passenger vehicle a hundred kilometres into space, twice within two weeks. Actually, X Prize did not, at that stage, have enough money for this prize. In 2002, Anousheh Ansari, a newly minted tech entrepreneur (who had dreamt of space flight since her childhood in Iran), and her brother-in-law, Amir, agreed to provide the funds and the prize became known as the Ansari X Prize.[7] It motivated twenty-five teams from seven nations to invest more than $100 million in pursuit of the $10 million purse. On 4 October 2004, the Ansari X Prize was won by Mojave Aerospace Ventures, who successfully completed the contest with their spacecraft, SpaceShipOne.[8]

---

5. Official Orteig Prize website.

6. Official Longitude Prize website.

7. 'How the Ansari X Prize Altered the Trajectory of Human Spaceflight', *Scientific American*, 4 October 2014.

8. Official Ansari X Prize website.

Amongst other interesting X Prize winners are Elastec/American Marine, which won $1 million in 2011 for recovering oil spilled at sea at a rate three times better than the industry's previous best (it has already launched a product based on this technology); and Edison 2, awarded $5 million in 2010 for a safe, cheap and easily built car that could do more than a hundred miles per American gallon (about 42 km/litre). The secret lay in an innovation: a novel system of suspension.[9]

The X Prizes are monetary rewards to incentivize three primary goals:

- Attract investment which is based on innovative approaches to difficult problems.
- Create significant results that are real and meaningful. Competitions have measurable goals, and are created to promote adoption of the innovation.
- Cross-national and disciplinary boundaries to encourage teams around the world to invest the intellectual and financial capital required to solve difficult challenges.

The idea of these prizes is, to quote Dr Diamandis, to set goals that are 'audacious but achievable'.

The X Prize Foundation has also launched the Google Lunar X Prize, sponsored by Google. With prizes totalling over $30 million, the challenge calls for privately funded teams to 'land a robot on the surface of the Moon, travel 500 metres over the lunar surface, and send images and data back to the Earth.' The first team to do this will receive a prize of US$ 20 million, while the second team to do so will get US$ 5 million.

---

9. This outline and some of the information in the next few paragraphs draws from a write-up in *The Economist*, 16–22 May 2015, besides various websites.

In November 2013, the X Prize organization announced that several milestone prizes would be awarded to teams for demonstrating key technologies prior to the actual mission. A total of US$ 5.25 million was awarded in 2015 for achieving the following milestones:[10]

- US$ 1 million each to three teams for the Lander System Milestone Prize to demonstrate hardware and software that enables a soft landing on the moon.
- US$ 500,000 each to three teams for the Mobility Subsystem Milestone Prize to demonstrate a mobility system that allows the craft to move 500 metres after landing.
- US$ 250,000 each to three teams for the Imaging Subsystem Milestone Prize for producing 'Mooncasts' consisting of high-quality images and video on the lunar surface.

There is only one team from India contesting for the Google Lunar X Prize. This is a group of youngsters—dynamic and technically qualified, but with no experience of space projects—called Team Indus. Their infectious enthusiasm has not only attracted more young engineers to their start-up, but also very experienced experts, many of whom are helping on a pro-bono basis. The competence of the team was recognized and their design validated when they were declared one of the winners of the three milestone prizes of US$ 1 million each. They are now working on the flight model of the moon lander, which they hope to have launched (possibly at the end of 2017) by ISRO.

The approach of defining an objective and encouraging

---

10. Official Lunar X Prize website.

organizations to compete to meet it has been used in the social sector too. An award of US$ 10 million has been announced for the winner of the Qualcomm Tricoder X Prize, inspired by the sci-fi handheld diagnostic device seen in *Star Trek*. The winner has to create a tool that can diagnose sixteen medical conditions, including anaemia, diabetes, strokes and urinary-tract infections, and is able to monitor five vital health signs, including blood pressure, respiratory rate and temperature. One of the contestants, Cloud DX, has already created a wearable device that captures these vital signs and then uses algorithms to spot potential problems. Another interesting prize is a US$ 15 million one for open-source software that will teach children in poor countries how to read, write and do sums within a period of eighteen months.[11]

Amongst others who have used prizes to spur innovation in the social sector is the Bill & Melinda Gates Foundation or the Gates Foundation. Its Grand Challenges in Global Health (GCGH) is a research initiative in search of solutions to health problems in the developing world. Fifteen challenges are categorized in groups among seven stated goals plus an eighth group for family health. The disciplines involved include immunology, microbiology, genetics, molecular biology and cellular biology, entomology, agricultural sciences, clinical sciences, epidemiology, population and behavioural sciences, ecology and evolutionary biology.

Bill Gates announced the Grand Challenges in Global Health at the World Economic Forum in Davos, Switzerland, in January 2003. In partnership with the US National Institutes of Health (NIH), the Gates Foundation granted $200 million

---

11. *The Economist*, 16–22 May 2015.

to the Foundation for the National Institutes of Health to establish and administer GCGH.[12]

Launched in 2008, Grand Challenges Explorations promotes inventions in global health research. Within three years, the Gates Foundation committed $100 million, and grants had been awarded to 495 researchers from 42 countries.[13]

In 2011, the Gates Foundation launched the Reinvent the Toilet Challenge—a programme created to design toilets that capture and process human waste without piped water, sewer or electrical connections, and transform waste into useful resources, such as energy and water, at an affordable price.

The Reinvent the Toilet Challenge is designed to spur innovation and bring creative thinking to solve the problem of dealing with human waste. By 2015, the Gates Foundation has funded sixteen research institutions across Africa, Asia, Europe and North America as part of the Reinvent the Toilet Challenge.

Meanwhile, in 2013, the Department of Biotechnology (DBT) under the Ministry of Science and Technology of the Government of India, and the Gates Foundation, in collaboration with India's Biotechnology Industry Research Assistance Council (BIRAC), launched a call for proposals as part of Grand Challenges India to reinvent the toilet, specifically in the Indian context. The Department of Biotechnology and the Gates Foundation will each invest US$ 1 million to support Indian investigators to drive research, development and production of the 'next generation toilet'.

---

12. Official Global Grand Challenges website.

13. 'Rewarding pioneering ideas: grants for innovative approaches to optimize immunization systems', World Health Organization, 6 September 2011; and Official Global Grand Challenges website.

Grand Challenges India supports co-funded projects to harness Indian innovation and research, and direct scientific discovery to develop affordable, sustainable solutions that improve health in India and around the world. As part of this partnership, the Achieving Healthy Growth through Agriculture and Nutrition programme was also launched in 2013, and seeks to target the relationship between agriculture, nutrition and health to reduce the high incidence of low birth weight and early stunting among Indian infants.

Apart from the Gates Foundation, the success of X Prize has generated many similar initiatives. The Methuselah Mouse Prize, created in 2003, offers cash to teams that breed longer living rodents, and thus contribute to knowledge about how animals age. In 2004, Bigelow Aerospace offered US$ 50 million to the first American team to create a reusable manned capsule that could visit a space station. Though the prize expired, unclaimed, in 2010, such a capsule has now been built (by SpaceX) and has flown, unmanned, six times.[14]

India's National Innovation Council (NInC)[15] too has sought to use challenge awards to spur innovation in desired areas. The National Innovation Council launched its first challenge through a call for proposals in October 2011 to reduce the drudgery of manual workers. It was an attempt to use scientific and technological capabilities to address problems of difficult manual work. The challenge was to provide decent working conditions for labour, including ideas that would improve the design of implements/tools, processes

14. *The Economist*, 16–22 May 2015.

15. These paragraphs on NInC are based on the author's personal involvement, as a member of NInC.

and models that improved working conditions, without displacing labour.

A total of 468 proposals were received of which thirty-three were shortlisted. These were examined by an expert group, which then invited six proposals for a detailed presentation. These proposals were from individuals, a group of students and a faculty team. NInC decided to award all six proposals, which included the design of a display unit for hawkers/street vendors, a cycle for disabled without hands and an innovative design for a rickshaw.

NInC also conceptualized the 'One MP–One Idea' annual competition. This sought to leverage the power of India's people through their elected representatives. The competition aimed to generate and select ideas by galvanizing all constituencies through the members of Parliament. The competition would invite innovative solutions in the areas of education and skills, health, water and sanitation, housing and infrastructure, agriculture, energy, environment, community and social services, etc. Submissions could be made by any individual, team or institution from within a constituency. The idea, proposed by NInC in 2011, was approved by the Lok Sabha but finally was not taken further.

Though not a targeted challenge (and with no award), another interesting initiative supported by NInC was Tod-Fod-Jod (break or take apart, and join). This involved taking apart a device (say a clock) and using the parts to make something else. Done extensively in workshops in schools, it was amazing to see the involvement of the students and the innovative things that they came up with. In 2016, the NITI Aayog announced a grant for setting up five hundred 'tinkering labs' in schools. The government will provide each school a one-time grant of

Rs 10 lakh and an equal amount over a period of five years for operational expenses. In addition, support will be provided to academic and non-academic institutions to establish a hundred 'Atal Incubation Centres'. This is part of the government's efforts to boost the innovation ecosystem in the country. The finance minister made an initial allocation of Rs 500 crore in the 2015–16 Budget, as part of the 'Atal Innovation Mission'.[16]

Challenge awards not only promote innovation; in many ways, this concept itself is an innovation. The prize or award might well be an attraction for innovators, but, more importantly, it is often the recognition and the opportunity to focus their work on a specific and defined problem that is the true reward.

Between the not-for-profit and the commercial sector, one key difference is individual motivation—in the former, many employees are likely to be driven by developmental goals. Despite this, the organizational factors that promote innovation are likely to be the same. In these organizations too, collaboration, communication, openness and minimal hierarchy are promoters of innovation, while bureaucracy, risk aversion and hierarchical command-and-control are inhibitors. Good educational institutions almost necessarily have the positive characteristics noted above, especially within their academic sub-structure. Little wonder, then, that it is the great universities (especially in the US and Europe) that have always been prolific generators of innovation, creativity and invention. It may also explain why Indian universities—with even those

---

16. 'Eye on startups, Niti to boost school labs', *The Times of India*, 27 May 2016.

considered the best being saddled with excessive bureaucracy and controls—have not been founts of innovations.

At the level of societies and countries, we noted earlier too that open, non-hierarchy-oriented, democratic and equitable communities are more likely to be truly innovative. Other factors, notably diversity and adversity, are also important drivers of innovation. Societal values and culture play a vital role, specially the attitudes towards wealth, risk and failure. A sense of national mission, induced by a degree of nationalism, can be a stimulant for problem-solving (and hence, innovation) as much as it is for hard work. One could argue that highly innovative Israel is an illustration of some of these—diversity (with its streams of immigrants from different cultures and countries), strong nationalism, democracy and adversity (resulting mainly from a back-to-the-wall, besieged-and-surrounded-by-enemies outlook).

India, on the other hand, has some of the factors in abundance (diversity and adversity, in particular), but also has inhibitors: the societal attitude towards failure and a yet-feudal, hierarchical social structure. As these change, innovation will flower.

It seems, then, that the factors that facilitate innovation in industry are not very different than those that do so in a not-for-profit developmental or educational organization, or in a country. Interestingly, the inhibitors or obstacles too are similar.

Given this understanding of the driving factors, what are the actions that organizations or countries can take to be more innovative? How can one move towards creating a society that is innovative? This is the subject of the next and concluding chapter.

8

A THOUSAND FLOWERS BLOOMING

Crafting An Innovative Society

Factors that create an innovative organization. Importance of interdisciplinary interaction. India's 3D advantage: democracy, diversity, demography. Steps to promote an innovative society.

Some feel that innovation, like its cousin, creativity, is a rare gift: a few have it within them and most do not. Others are of the view that innovation is an inherent part of the human genome, that all of us have it in-built, even if in varying degrees. Thinking—'I think, therefore I am'—is what defines human beings, and leads naturally to innovation. The problem is that the path from thinking to innovation has many obstacles. Even when these are overcome, the manifestation of the innovative idea is often constrained by fear of failure, discouragement or derision ('what a stupid idea'; 'that will never work'). As a result, innovation is generally uncommon.

Between the two schools of thought, we subscribe to the latter and take the view that innovation is innate, a basic trait common to all humans. This premise naturally poses the challenge: how best to bring out the latent innovation and how to develop or stimulate the innovative capabilities of individuals. Earlier chapters (Chapter 5, in particular) have identified some factors that are conducive to innovation and promote it. Can organizations or countries capitalize on these? While some of these factors may seem to be immutable givens in certain contexts, the view taken here is that appropriate interventions through policies, structures and processes can amplify—possibly even create—these factors. Thus, it is possible to make countries and organizations more innovative. What follows looks at some of the major factors that drive innovation and suggests what might be done in each case.

As mentioned frequently in this book, diversity is amongst the important drivers of innovation. Fortunately, this is something that is not difficult for organizations to create. In fact, either out of genuine commitment or to be politically correct, many organizations—in many countries—are already laying a great deal of emphasis on ensuring diversity in their employee base. For most, this means working towards a better gender balance by promoting the recruitment of women. To many, it entails the recruitment of the differently abled ('persons with disabilities', in official Indian terminology). In India, government organizations also have a quota or reserved positions for those who are from the categories of Scheduled Tribes, Scheduled Castes and Other Backward Classes. Like other efforts, this compulsory caste-based reservation does ensure greater diversity. A few organizations—preponderantly those with multi-country operations—try to recruit nationals from different countries, thereby seeking to enhance the diversity of nationalities in their employee base.

The diversity that these efforts aim to create is a goal that is worthwhile in itself. However, in addition to this, if the purpose is to stimulate innovation, then another dimension of diversity is needed. In addition to gender, physical ability, caste/class and nationality, what is required is diversity of thinking. Part of this is a consequence of upbringing and experiences, and a fair range of these is automatically ensured through the diversity factors mentioned earlier. Yet, it is better and more effective to focus specifically on recruiting people with 'different' thinking. This requires that one moves beyond the recruitment processes that result in the selection of standardized, best-fit people, and consciously look for those who think out-of-the-box and tend to come up with unconventional ideas and solutions. This will need not only an overhaul of the methodology of recruitment and assessment, but also a change in the mindset of the human resources team. The starting point, through, has to be at the top management level. The organization must have an explicit policy of recruiting people with thought diversity, just as it has a policy regarding, for example, ensuring gender diversity.

In an earlier chapter, other steps to stimulate, foster and promote innovation in organizations have been discussed. In summary, these include:

1. A work culture that encourages openness, discussion and dialogue across horizontal and vertical boundaries in the organization.
2. Physical design and seating that enhances interaction amongst employees, again across levels and departments. This must include sufficient and appropriately located leisure and utility spaces like cafeterias, pantries, water dispensers, rest rooms, reading rooms, etc.
3. Putting in place human resources policies and recruitment methods (particularly screening, tests and interviews)

that ensure as much diversity as possible in the employee base.

4. Organizational showcasing and recognition for the best innovations, based on proactive promotion through open-ended or focused contests. Equally important, recognition and awards for 'grand failures'—potentially path-breaking ideas that, for some reason, did not succeed.

5. Creating an innovation group, freed from the regular and routine organizational functions, which is given the time, space and resources to work on generating and furthering new ideas. Ideally, employees should be rotated regularly through this group, so that those with an understanding of real problems are the ones seeking innovative solutions. This group could take the form of an in-house incubator, catering to 'intrapreneurs' and providing the kind of support and facilities that incubators give to entrepreneurs.

6. An alternative, or supplement, to (5) above is to allow every employee free time to work on projects and ideas that appeal to him/her.

7. Ideas and innovations are often catalysed by input from outside, sometimes from a completely different field. This could be done by having regular talks, or even occasional workshops, on topics far removed from the work of the organization. In large organizations working in different fields, and conglomerates with diverse products, encouraging and facilitating interaction across disciplines within the organization can be a substitute for seeding ideas through an outside input.

8. Providing employees funding and leave to become

entrepreneurs to take forward their innovation and, at an appropriate time, reintegrating the innovation and employee back into the organization.

9. Organizing creativity workshops on a regular basis, and arranging employee visits to innovative companies. Inviting successful innovators to give talks, sharing their experiences.

10. Sending a clear message from the top (from the level of the CEO) that the organization greatly values innovation and innovators. This should be backed up with concrete actions—rewards and recognition—which convey that the message is real and serious.

These suggestions are equally applicable to commercial and not-for-profit organizations, including academic institutions. Obviously, the specific context would require modifications in some or all the suggested actions. Also, these would be broadly applicable, irrespective of the size of the organization. The idea is to create, within the organization, an ecosystem that facilitates innovation.

As one moves to the broader level of a country, are there steps that can be taken to facilitate innovation? This is a crucial issue in light of the growing importance of innovation, and the desire to create an innovative society. From such a society will spring individual innovation, which is key to the progress of the country and the growth of enterprises. Earlier chapters have touched on various steps that may help. Some actions that would contribute are outlined in the following paragraphs.

Making a whole society innovative is an evolutionary and long-term process. In a risk-averse milieu, doing something new entails a high social cost. Also, in a conservative and tradition-bound society, being different (and, by extension, even thinking differently) is looked at askance. Rigid hierarchy is another show-stopper as noted in the case of organizations. This is true of societies too. Much of Indian society has just these characteristics: risk aversion, conservatism, traditionalism and hierarchy. If there is yet a fair degree of innovation, it is in spite of these factors.

Overcoming these requires major social reform and change. One part of the change must be at the very base: in the school system. Conventional schooling prescribes that, while in the classroom, children must keep their eyes and ears open, and mouth tightly shut (no talking, no questions). In the examination, they are expected to regurgitate whatever is in the textbook, or the teacher's words. This has to change. Questions and discussions are an essential part of learning, and promote independent thinking. The latter is also a necessary condition for innovation.

Conventional schooling is individualistic; all assessments are of individual performance, and teamwork or group projects are a rarity. This, too, has to change, since innovation often results from teamwork, and taking an idea forward inevitably requires group work.

Almost all scholarships and recognition are based an academic performance. If innovation is to be promoted, awards for it are necessary. There has been a suggestion to set up National Innovation Fellowships for children in the thirteen- to sixteen-year age group. The concept is to base these on a nationwide contest, with applicants required to work in teams and under the mentorship of a teacher. The award will be for the selected teams, as also for the mentor-teachers of these teams. The idea is to not only recognize and foster innovation amongst youngsters, but also to incentivize teachers to promote innovativeness. The fellowship will consist of a stipend for the team (and the teacher) for two or three years, enabling them to work on further developing the innovation, and will include workshop sessions where they can meet other innovators and receive guidance from experts.

Such an approach, beginning at the level of schoolchildren, is obviously a slow and long-term one, but is likely to have a large-scale and sustainable impact. This is also a difficult step, since it requires a radical change in the teaching-learning model and in the examination method, deviating from a long-entrenched system.

At the tertiary level, the trend in India over the last two decades has been to set up single-discipline universities—for example, in IT, law, fashion design, humanities, medicine, management, etc. Such unidimensional universities are, in fact, a travesty of the meaning of 'university' (which derives from the Latin *universitas* meaning 'a whole' and the Old French *universite* meaning 'universality'[1]), and are unable to provide the opportunities for interdisciplinary work (or even interaction) that conventional universities do. To the extent that innovation flowers best when there is cross-pollination amongst disciplines, the specialized universities are not conducive to innovation. The government needs to reverse its policy in this regard and embed specialized institutes within broad-spectrum universities. It must also encourage the creation of multidisciplinary centres in these universities, so that there is a platform for interaction amongst diverse streams.

A recent example of how innovative thinking is a result

---

1. The Oxford English Dictionary; Online Etymology Dictionary.

of cross-fertilization amongst disciplines is the new work on drones.[2] These pilotless vehicles are now transiting to civilian and practical use, for example in aerial photography, surveillance and delivery of goods. In March 2015, Amazon, the online US retailer, was given permission to test a drone to actually deliver its merchandise. Ultimately, these drones will need to fly autonomously (i.e. without an operator on the ground to the guide them). A combination of the satellite-based Global Positioning System and Google Earth can help a micro-drone navigate around permanent obstacles on its route, by pre-programming its course. The problem, though, is the unexpected—unmapped trees, gusts of wind or even an unwary bird. Building senses into the drone to deal with these is obviously essential, and this is where inter-disciplinarity steps in.

Laboratories around the world are now working on this by looking at how the original drones (the insect variety) fly, and moths manage to avoid sudden obstacles. From aeronautics and electronics, the researchers now have to understand biology, if their drones are to function effectively in the real world. For example, Dr Ashutosh Nataraj from Oxford University drew inspiration from a bee in order to give his drones vision. Trying to understand how the bee avoided obstacles in a human dwelling, he studied apidological literature and picked up ideas from that for his drones.

The importance of learning from the animal world (and thus of cross-disciplinary flow of ideas) is highlighted by the fact that, adjusted for size, blowflies are better at manoeuvring

---

2. Information on the work mentioned in these paragraphs is available in various places, including *The Economist*, 28 March 2015.

than the latest fighter aircraft. This requires a combination of vision and inertial guidance. Several research groups are now studying insect inertial guidance. Interestingly, in contrast to conventional engineering, flies do not pass the sensory input to their brains for processing; instead, the signals act directly as a series of reflexes controlling speed, heading and attitude.

Two other groups, working at the Johns Hopkins University (led by Rajat Mittal and Noah Cowan), are studying hawk moths. They found that moths held their heads and thoraxes steady with respect to a flower while hovering over it by making minute changes to the orientation of their abdomens. Dr Cowan has used this finding to modify a drone so as to stabilize it. At Harvard, Robert Wood has taken a different approach to the problem of hovering—he has created artificial 'eyes' for micro-drones, based on ocelli. These are small eyespots that insects use to take bearings of the sun or the moon, enabling them to maintain a straight course by keeping a constant angle to the distant light sources.

These new studies and ideas that will, doubtless, make possible the early practical uses of micro-drones, highlight the innovations that are facilitated by developing linkages amongst disciplines. What better way to facilitate such cross-disciplinary interaction than to have universities that cover diverse subjects and promote an active intermingling of researchers?

As discussed earlier, organizations could promote diversity by appropriately modifying their recruitment norms and processes. Similarly, educational institutions too need to re-examine their admission mechanisms, so as to ensure a diverse student body. Given the paucity of truly high-quality university-level institutions, the number of applicants for a limited number of seats is huge. For management courses, the ratio for the top IIMs is near 100:1, which means that one has to be at the 99 percentile level (in the top 1 per cent) to secure a seat. The situation with regard to engineering (for admission to the top IITs is similar. The result of this massive demand is that admission is based on purely objective tests, performance in which can be considerably enhanced (though some disagree) through coaching classes.

A consequence of the standardized filter (namely, the tests) and intensive coaching, often from an early age, is that those admitted tend to be similar in the way they approach a problem. Many years ago, after talking to hundreds of those admitted to the top professional institutes, I classified it as the 'Mother Teresa Syndrome'.[3] This originates from the fact that when, several years ago, Indian women were sweeping international beauty contest awards, it seemed that the standard (and, of course, much-appreciated) answer to that final decider question (irrespective of whether the question was 'who is your role model' or 'who is the greatest living person' or anything else) was 'Mother Teresa'! In a similar manner, when asked who their role model was, eight out of ten of the newly admitted students would say 'President Kalam'. He was, undoubtedly, a great person, and someone whom young people revered. Yet, it was surprising that so few named a great scientist, writer, political leader or sportsperson. Clearly, either this was their genuine response (which indicates that the selection process was filtering out any diversity) or the coaching classes had, in effect, brainwashed everyone to give the same 'desirable' answer. In both cases, diversity was eliminated.

---

3. 'Role models: Build on inherent diversity for innovation', *The Economic Times*, 19 November 2007.

Contrast this with the admissions process to the top US universities. They too have standardized tests (SAT, GRE, GMAT), but these are supplemented by many essays to be written by the applicant. Assessing the essays requires a great deal of effort and is certainly subjective, but enables considerable diversity to be brought in. In India, on the other hand, the weightage for interviews or group discussions is being reduced. Recently, it was decided to abolish interviews for recruitment to certain government positions. We seem to confuse subjectivity with bias. The latter is to be avoided and condemned; the former is often necessary. Lack of trust—based, no doubt, on many instances of its abuse—is the core problem.

Another reason is the bureaucratic preference for standardization and centralization. Today, despite the objective Common Admission Test, each IIM gives different weightages to different sections of the test and then prepares its own shortlist. However, there is now pressure to move towards a common all-IIMs ranking (as the IITs do). This will remove even the small modicum of diversity among the students of the various IIMs.

This standardized and centralized filtering may well be one reason for the comparative lack of path-breaking innovation or research and development in our universities. Yet, India's advantage of inherent diversity (on so many dimensions: ethnicity, language, religion, income, social environment, culture, etc.) probably ensures that some degree of divergent thinking remains even amongst the seemingly cloned students. If, in addition, one were to consciously promote a more diverse student body, innovation may well take off like a rocket!

The importance of diversity as a catalyst of innovation has

been mentioned earlier, as also the fact that India has a natural advantage in this respect. It is essential that this diversity be promoted and celebrated. Now and again there is a push towards an attempted homogenization, with an excessive focus on commonality and the concept of Indian culture (in the singular).

The drive towards enforcing uniformity, rather than the concept of unity in diversity, is worrisome on many counts. From the limited viewpoint of innovation, it is counter-productive and will negate India's biggest advantage in the innovation sweepstakes. Just from this point of view, government policy must proactively promote diversity. Further, as in the case of different academic disciplines, the value lies in interaction. Therefore, to reap the benefits of diversity, platforms and institutions must bring people with diverse backgrounds together. It is the intermingling of these diversities that acts as a spur for innovation.

Democracy is another existing advantage for India. Free discussion, debate and openness were noted earlier as important factors for innovation within organizations. At a societal level, a democracy provides this. It is essential that this element of democracy be recognized and furthered. Too often democracy is seen merely as the right to elect one's representatives through a free and fair election once every few years. However, for innovation to flower (and, obviously, for a host of other larger reasons), it is necessary that other elements of true democracy be given importance, most of all, vigorous and free debate.

Another key facilitator of innovation, which is an integral part of democracy, is equality. Hierarchy and a command-and-control structure may possibly aid efficiency, but it is anathema for innovation. A society cannot achieve its full innovative potential if it works on edicts, with no freedom to debate and discuss on an equal basis. Democracy provides these, and innovation demands such equality as a basis for its success.

One final and key element of democracy that is crucial for innovation is the right to dissent. Its importance is obvious, given that the genesis of innovation is a thought or idea that differs from the present practice. Providing space and protection to dissenters is a vital part of democracy and one that is, unfortunately, under threat. Too often democracy degenerates into majoritarianism, and differing viewpoints are delegitimized. Even in a vibrant democracy like India, there are too many attempts to stifle dissent. Censorship, bans and labels of 'anti-national' are becoming more common, even as the pressure to conform—exerted through community organization like khaps or by moral policing—is increasing. This regression is undesirable and can be a major obstacle to innovation.

The protection and promotion of diversity and democracy is a necessary part of building on India's 3D advantage (the third D being demography) for innovation. The higher proportion of the young means that there is a freshness and vibrancy in society, with new perspectives, questions and ideas coming together with the energy and dynamism of youth. Because of their natural curiosity, willingness to explore and rebelliousness, it is inevitably the young who come up with new, often crazy, ideas. Within an appropriate milieu, this is the foundation and genesis of innovation. Therefore, a country with its population having a median age of twenty-seven years (in 2014) is indeed well placed to be an innovative nation. This contrasts with a median age of 37.6 years in the US and as much as 46.1 years in both Japan and Germany. Interestingly, China (because of its one-child policy) is also faced with a rapidly aging population, the median age of which is already 36.7 years.[4] According to one projection, as much as 39 per cent of China's population will be above the age of sixty by 2050.

The tremendous diversity of India is, as noted, a potential trigger of innovation. Within the country, cities like Mumbai, Bengaluru and Delhi/NCR have the most diversity. Little wonder, then, that they are also the hubs of innovation (see Chapter 6). Cities aspiring to become centres of innovation must see how they can increase their diversity index. This entails taking steps to make the city attractive to outsiders. The first step must be to reduce—if not eliminate—insularity, parochialism and rigid and conservative social values. The local administration and the concerned state government have a major role to play in this, and they need to work in partnership with progressive civil society organizations. So far, this has not been done; in fact, there are cases where the policies and actions of the state government have been regressive. For states seeking to attract entrepreneurs and become innovative, this presents an opportunity. In addition, proactive actions to quickly develop the necessary physical infrastructure (Chapter 5 has outlined this) will add to the attractiveness.

The diversity of the country as a whole lacks in one dimension, and on this India is far behind many others: diversity of nationality. This aspect enhances the range of diversity, especially when one considers those from more distant (not merely neighbouring) countries. India needs to increase its population of expatriates and this means that the

---

4. Data from 'Statista—The Statistics Portal'.

country must have a policy that encourages talent from around the world to come to India. Many countries, including the US, Singapore and Israel, welcome qualified immigrants. The US tech industry is widely perceived as being powered by H-1B visa-holders, many of whom transition to residents ('green card' holders) and on to citizenship. India's present policies and procedures are an obstacle for foreigners wishing to work in the country, and need to be radically changed. At present, obtaining visas for foreigners invited to a conference is a painful process, leave alone getting clearances for someone who wants to work in the country for a few years. A change in mindset, policies and processes is essential.

Surveys continue to rank India poorly in terms of its attractiveness for expats. Certainly, it is exciting to foreigners in terms of opportunity, potential and dynamism; and yet, it is a rather challenging place to live in. A survey by InterNations,[5] the world's largest social network for expats, put India in the bottom ten (in both 2014 and 2015), being ranked 55 of 64 countries surveyed. Singapore (at number 4) and Thailand (7) far outrank it. Amongst the top three concerns are personal safety/crime, cost of living and climate/weather. Apart from these, one of the issues seems to be work–life balance, as expats working full-time in India put in an average of 46.5 hours per week.

Despite the difficulties, the market opportunities and cultural aspects of India seem to attract foreigners, many of whom are willing to tackle the various obstacles and set up a business in India. In discussing the missed-call gambit, we

---

5. 'The World Through Expat Eyes', *Expat Insider*, The Inter Nations Survey, 2015.

mentioned ZipDial and its American entrepreneur, Valerie Wagoner (Chapter 2). Of late, there have been many more expats setting up ventures in India. A recent article[6] covered an interesting sampling of these: from South Africa, Australia, Italy, France, Canada, the UK, Germany and the US, with a wide range of businesses located in various cities across India.

As seen from experience elsewhere, the best starting point to promote diversity of nationality is to attract students. Inevitably, some of them will want to stay on and work in the country. Equally important, a diverse student body (as also faculty) greatly enriches the interaction and provides a stimulus to innovation. India is well placed to attract foreign students, given the recognition enjoyed by its top institutions (even though these are very few) and the lower cost of education. However, there is a need to substantially upgrade the physical facilities in the universities. Students (even Indians students) are no longer satisfied with spartan facilities in hostels or classrooms. The days of universities epitomizing 'simple living and high thinking' are long gone. Air-conditioning, good and varied cuisines, state-of-the-art classrooms and high-speed Internet are all expected. These are, therefore, necessary conditions to attract global students. Facilities and Chair professorships for foreign faculty, and scholarships for students, are well worth providing; these will be more than compensated for by the enrichment (including innovation) that such diversity brings. While some of the actions suggested (visa policy, for example) fall in the domain of the central government, there are many things that can be done by the states. Those who do act will be one step ahead in the innovation sweepstakes amongst states.

---

6. *The Economic Times Magazine*, 24–30 May 2015.

The value of interaction between academia or research institutions on the one hand, and industry on the other, is well recognized. The closeness and intensity of interaction directly impacts the extent of innovation. While many factors determine this, physical proximity is certainly an important one (despite the obliteration of distance by new technologies). It is essential that this be kept in mind when new facilities are being created. The synergies and advantages of clustering are known, and planning needs to take note of this. State and city authorities could earmark land for such knowledge zones and provide incentives (lower costs, for example) to attract educational/ research institutions and related industry and incubators to set up facilities in these zones. Since the setting up of educational or research institutions is an incremental process, it should be government policy to locate new institutions in proximity to existing ones, to the extent possible. In India, collaboration between academic/research institutions and industry has been limited. To give this a boost, it may be worth thinking of special incentives—for example, special grants and low-interest loans for such joint work, or even tax write-offs for expenditures involved in such joint work.

Interaction between academics/researchers and industry/ entrepreneurs could also be encouraged by permitting a faculty member or researcher to own equity in companies spawned by their work and inputs. Many institutions already permit this, but it needs to be universalized. A more direct research-to-entrepreneurship route exists and must be encouraged. This entails the researcher himself/herself founding the enterprise. This, as experience has shown, is fraught with some drawbacks. One of the problems is that researchers do not necessarily have good business sense. Therefore, they are better suited to being mentors or technology advisors—possibly even the chief technology officer—of the venture, rather than being the CEO. However, the direct involvement of the researcher in a venture, in terms of time and commitment, can add great value, especially in solving the technology problems in the laboratory-to-market journey. It is, therefore, well worth encouraging researchers to become part of the venture that their work has triggered. This could be done by permitting them to take up to two or three years' leave (in the Indian context, one year is unlikely to be enough for a start-up to find its feet and test the market) and also providing seed capital for such ventures. Some institutions have a provision for leave, but this needs to be more widespread. In addition, there is a need to set up a special fund to support such researcher-initiated ventures. Also, the institution must allow the researcher to continue using the laboratory and other facilities while on leave.

These steps would not only ensure the practical use and monetization of research, but could also trigger an entrepreneurial surge. More importantly, it would encourage innovation on a wide scale by providing an incentive for research and through the process of converting ideas and concepts into marketable products.

Another initiative could be the creation of co-working facilities, where start-up ventures can begin their journey. Very often, all that is needed is fitted-out space with high-speed Internet connectivity, conference rooms and a cafeteria. The value comes mainly from the interaction amongst the entrepreneurs. Rapid prototyping and testing facilities, available on a pay-per-use basis, would greatly help product companies. Entrepreneurs should be encouraged to set these up and where necessary, incentives to do so could be provided (through, for example, subsidized space). Alternatively, research institutions or universities could provide these services.

Major technological innovations—those that have a significant and long-term impact—inevitably require substantial investment in research and development. Much of this must necessarily come from the government. Yet, given the potential benefits that could flow from a well-directed research and development programme, industry too should be making substantial investments in research. In fact, many companies actually do so, especially in the rapidly changing arena of technology. Cisco (with a turnover of US$ 49.2 billion in FY 2015) invested as much as US$ 6.2 billion (12.6 per cent of its revenue) in research and development.[7] Volkswagen spends more on research and development than any other company on the planet: €13.1 billion in 2014.[8] It continued to be the global leader in research and development spend in 2015 (US$ 15.3 billion).[9] Indian industry, barring a handful of companies, has invested little in research and development. Even the highly profitable IT sector has a disappointing record in this regard. While there have been occasional tax incentives to industry to promote investment in research and development, an across-the-board (instead of sector-wise) incentive for a sustained period (rather than year by year) may serve as a good stimulus. In this respect, it is welcome that research and development is included in the approved list for corporate social responsibility expenditures (mandated to be 2 per cent of the average of profits over the last three years, for certain categories of companies, in a comply-or-explain mode). However, if innovation is to be given a sustainable long-term

---

7. Cisco Systems, Inc. 2015 Annual Report.

8. 'A mucky business', *The Economist*, 26 September–2 October 2015.

9. 'The Global Innovation 1000: Top 20 R&D Spenders 2005-2015', PriceWaterhouseCoopers.

boost, it is essential that the government increase its investment in research and development. India's record in this has been dismal: against the long-announced target (somewhat modest, compared to the leaders) of 2 per cent of GDP for research and development, government expenditure on the same hovers around 1 per cent, and there has been little visible progress in moving to 2 per cent. To create an innovative society, the government needs to make much larger investments in research and development.

The value of dissent—of thoughts, ideas and views that are different from those commonly held—as a catalyst for innovation has been discussed earlier. In creating an innovative society, few factors are as important as this. A conscious policy to protect, even promote dissent must be an integral and necessary part of any effort to foster large-scale innovation and to embed it as one of the core attributes of a society. Academia and civil society are organized groups that often espouse causes and ideas that are contrary to mainstream or conventional thinking. Their freedom to do so must be protected and nurtured.

Building an innovative society is a long-term proposition, with the effort beginning at the school level itself. Changes in the education system have been discussed earlier, as also the suggestion for a National Innovation Fellowship. These and other initiatives could be taken by state governments, given that school education is within the ambit of responsibilities of the states. Another initiative, already discussed, is the idea of challenge prizes. This concept has been successful elsewhere, and India too is now adopting it. This can be taken up by industry, states or by the central government, or even through a joint effort amongst them. Apart from focusing creative energies on a specified problem, the visibility and buzz around such grand challenge efforts will help to encourage innovation across the board, creating a stimulus for an innovative society.

Clearly, as outlined in this chapter, there are many actions that the government (both, central/federal and state/regional) can take to promote innovation. Any country that desires to be at the forefront of economic progress in the years to come must ensure that it is entrepreneurial, creative and innovative. The government, through judicious and proactive intervention, can play an important role in this. Industry, through individual companies acting in enlightened self-interest, can also be a major stimulator of innovation. Industry and government, working together with academia, can ensure synergies through partnerships. This is a necessary step towards developing an innovative society.

Creating an innovative society is an ambitious and challenging goal. It is something that many countries speak about and aim for, and a few are taking concrete steps to achieve. The task is necessarily one that demands focus, determination and a long-term perspective. It requires a combination of national policies, specific government actions, and many steps by organizations; it also needs collaborative work among the government, industry, academia and civil society organizations.

For a society to be innovative, it must have a high degree of innovation at the level of the individual. Organizations must foster and encourage innovation, and have appropriate mechanisms to channel individual or group effort into overall organizational innovation. Universities and research institutions are also 'organizations', but given their special position, must be leaders of innovation. Interaction between industry and academia should catalyse and synergize even greater innovativeness.

Civil society has a special role. It should help to identify and support innovation at the grass-roots level, especially in rural and remote areas. It must also support diversity, particularly of thought, at a time when conformity is often expected and demanded. Often, such 'different' thinking is dissent, which is sometimes seen as being undesirable, even anti-national. Yet dissent, apart from being the foundation of democracy, is the seat from which innovation grows and flowers. It falls mainly on civil-society organizations to ensure that dissent is not stifled or discouraged.

In India, as in many developing countries, given the ubiquitous presence and extensive involvement of government in practically all activities, its role in promoting innovation is of vital importance. It is the policies of government (central and state) that can be key facilitators and promoters of a society that is innovative. Apart from overall policy, specific actions can play a catalytic role. Investment in research and development, challenge contests, funds and facilities for innovation-based entrepreneurial enterprises, innovation fellowships for youngsters—these and a host of other initiatives can help to drive innovation amongst various segments of the population. Such efforts across a cross-section of people are essential if one is to develop an innovative society.

Many factors drive innovation, but in India, as noted earlier, adversity has been an important one. However, typically, adversity triggers improvisation or jugaad, rather than the more impactful and sustainable form of innovation. Therefore, without in any way downplaying the role of improvisation, one has to also think of ways of promoting a fuller form of innovation. This is also important from a long-term point of view: as India develops, one hopes, and expects, that adversity decreases. It will then be necessary to look for and amplify other stimulants of innovation, as also to promote beyond-improvisation innovation.

India faces another challenge, one that is uncommon elsewhere in the world, if not unique to this country. This is the task of tapping and channelizing the creativity of system-beating or system-bypassing innovators into positive and system-enhancing innovations. Excessive discipline or straitjacketing may well dampen the innovative urges; on the other hand, near-anarchic situations (road traffic, for example) tend to encourage zero-sum, me-first innovations, which are detrimental to society. New and innovative solutions need to be found to this conundrum. After all, unlike pure creativity, innovation must provide actual or potential societal benefit.

The government is now looking at entrepreneurship and start-up ventures for societal benefits such as creation of wealth and jobs. Prime Minister Narendra Modi, in his Independence Day speech on 15 August 2015, announced the launch of Start-up India, Stand-up India. The government has also indicated that it will ease procedures for setting up start-ups and provide a far more conducive tax regime for them. The prime minister personally launched Start-up India on 16 January 2016. This is an excellent initiative, and intended to complement and supplement the manufacturing thrust of the Make in India programme. It may be good to backward integrate this so that the slogan can be 'Imagine and Innovate in India'.

The dream of an innovative society—one in which innovation takes place regularly, even routinely, in classroom and office, farm and factory, at home or on the road—is a realizable one. Countries will compete to be more innovative, and the future will belong to the winners. As each country seeks to better others, innovation will spread. And the world will become a better place for all.